50 STEM SUPERJOBS

YOUR FUTURE IN SCIENCE, TECH, ENGINEERING AND MATH

TABLE OF CONTENTS

PART 4: MATH MAGICIANS

Statistician – Truth Finder in Numbers
Actuary – Risk Calculator
Operations Research Analyst – Big Problem Puzzle Solver
Cryptographer – Code Maker & Code Breaker
Financial Quant – Math for Markets
Epidemiologist – Disease Pattern Tracker
Logistics Planner – Supply & Delivery Wizard
Game Theorist – Strategy Scientist
Math Communicator / Teacher – Makes Math Make Sense
Astronomer – Universe Number Explorer

PART 5: PLANET PROTECTORS & SPACE EXPLORERS

Renewable Energy Engineer – Solar & Wind Hero
Conservation Biologist – Habitat Protector
Urban Planner – Designs Smart Cities
Agricultural Scientist – Future Food Grower
Disaster Scientist / Engineer – Storm & Earthquake Safety
Space Engineer (Mission Specialist) – Rockets & Rovers
Astrobiologist – Looks for Life Beyond Earth
Ocean Engineer – Underwater Tech Builder
Geoengineer / Hydrologist – Earth & Water Expert
STEM Entrepreneur – Builds Their Own STEM Company

INTRODUCTION

STEM Is Not Just for Geniuses (It's for You)

When you hear the word STEM, what do you imagine?

Super hard math problems?
People in lab coats mixing strange liquids?
Kids who were "born smart" and always know the answer?

Here's something important you should know right away:
STEM is not just for geniuses.
STEM is for kids who are curious.

STEM is for kids who ask questions.
Kids who like building, fixing, drawing, coding, organizing, or imagining.
Kids who care about animals, the planet, space, games, people, or problems that don't seem fair.

STEM stands for:

- Science – asking questions about how the world works
- Technology – using tools to make life easier or better
- Engineering – building solutions that actually work
- Math – the language that helps us measure, compare, and understand

But really?
STEM is **curiosity + creativity + problem-solving.**
You don't have to love every part of STEM.
Most people don't.

Maybe your brain lights up when you:

- Take things apart to see what's inside
- Organize information and notice patterns
- Imagine new worlds and how they would function
- Worry about animals, pollution, or the future
- Enjoy teaching others how to do something

Those are not random habits.
Those are STEM starter superpowers.
This book is a tour of 50 real STEM careers that exist right now—and a few that are growing fast as the world changes.

For each one, you'll discover:

- What they really do (in kid-friendly language)
- Where you can see their work in everyday life
- Cool tools and technology they use
- Big future problems, they help solve
- A simple activity you can try yourself
- What kind of person might enjoy this kind of work?

You do not have to pick your job today.
Your job right now is to:

- Learn
- Explore
- Try lots of things
- Pay attention to what feels exciting or meaningful

Think of this book like a **career museum.**
You're just walking around, looking at what exists, and noticing which rooms make you want to stop, explore, or come back again.

Some careers will sound amazing.
Some will sound boring.
Some might surprise you.

That's all normal.
By the end of this book, you don't need an answer.
You just need better questions.
Ready?

Let's meet 50 STEM superjobs—and maybe discover a few clues about future-you along the way.

PART 1

Science Detectives

Finding Answers by Asking Better Questions

Some people think science is about knowing the right answers.
It's not.

Science starts with noticing things that don't quite make sense and asking questions about them.
Why did that happen?
What's really going on here?
How could we find out?

Science detectives are the people who look closely at the world and try to understand it. They study tiny things you can't see, huge systems like oceans and climate, and even what's happening inside your own brain and body.

They don't guess.
They investigate.

As you explore these careers, pay attention to what catches your curiosity. You don't need to understand everything. You just need to notice which questions make you want to learn more.

That's how every science detective begins.

Microbiologist – Germ Hunter

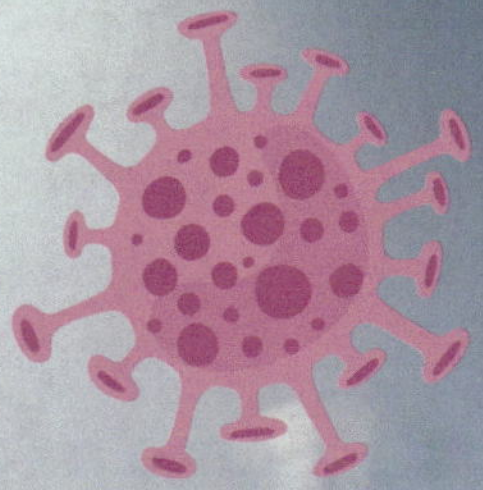

Imagine This...

There's a whole world around you that you can't see.

It's in your hands.
On your desk.
In the air.
Inside your body.

Most people never think about it

But you do.

You wonder what's living there, what it's doing, and whether it's helping you—or trying to make you sick.

If tiny, invisible things make you curious
instead of grossed out, you might already be thinking like a **microbiologist.**

What They Do

Microbiologists study microbes—living things that are too small to see without special tools.

These include:

- Bacteria
- Viruses
- Fungi
- Other microscopic organisms

Some microbes can make us sick.

But here's an important truth:

👉 Most microbes are actually helpful.

Microbiologists try to figure out:

- Which microbes cause illness
- Which ones help our bodies work properly
- How microbes spread from place to place
- How to stop harmful microbes without destroying helpful ones

They are detectives in an invisible world, using clues most people never notice.

Where You See Their Work in Real Life

You see the work of microbiologists more often than you realize.

You see it when:

- A doctor prescribes antibiotics
- You eat yogurt with "live cultures"
- Schools remind students to wash their hands
- Scientists talk about vaccines or outbreaks

Microbiologists also help:

- Keep drinking water safe
- Make food safer to eat
- Protect patients in hospitals
- Help farmers grow healthier crops

Even though you can't see microbes, microbiologists help make everyday life safer.

Cool Tools & Tech

Microbiologists use tools that let them explore the invisible world, such as:

- Powerful microscopes to see tiny cells
- Petri dishes where microbes grow in patterns and colors
- Incubators that create warm environments for microbes
- Pipettes for moving tiny drops of liquid
- Computers to track how germs spread

Wow fact:

👉 A single drop of water can contain more living organisms than there are people in your entire city.

Future Problem They Help Solve

Microbiologists help solve some of the world's biggest challenges, including:

- Preventing future pandemics
- Creating new medicines when old ones stop working
- Finding microbes that can clean pollution or break down plastic
- Protecting people with weak immune systems
-

Some even study microbes that live in volcanoes, deep oceans, or freezing places to understand how life survives extreme conditions.

Try It Now: Germ Detective (Observe)

You'll need:

- Your hands
- Soap and water

Steps:

1. Look at your hands. They look clean, right?
2. Think: Where might tiny living things be hiding?
3. Wash your hands with soap.
4. Ask: Why do scientists still care about germs we can't see?

👉 This is how microbiologists think—about invisible life.

You'd Like This If You...

- I am curious about germs and sickness
- Don't mind careful rules, gloves, or goggles
- Like science that helps people stay healthy
- Enjoy noticing small details
- Think "ew" things are secretly fascinating

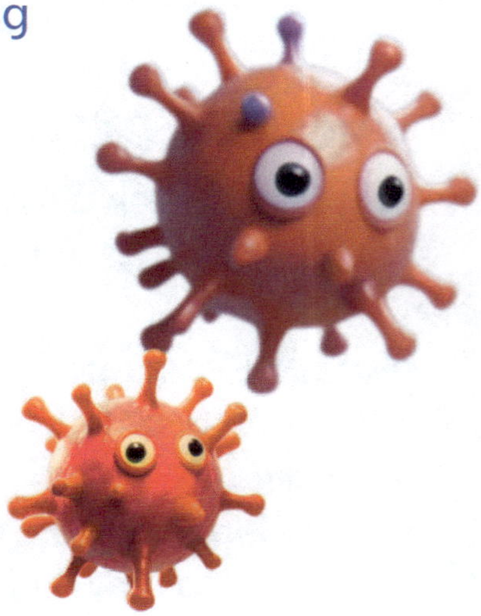

Why this matters to you:

Understanding germs helps you make smarter choices about your own health and helps protect the people around you, too.

Microbiologists don't just fight germs. They help us understand the invisible world that keeps life on Earth going.

And once you start noticing it, you'll never look at a doorknob or your own hands the same way again.

Marine Biologist
Ocean Creature Guardian

Imagine This...

You're standing near the ocean.
Waves move in and out.
Seagulls call.
Something splashes just below the surface.
Most people see water and keep walking.
You wonder what's living under there.
Who swims below the waves?
How do they survive?
And what happens when their ocean home starts to change?
If the sea makes you curious instead of calm, you might already be thinking like a **marine biologist.**

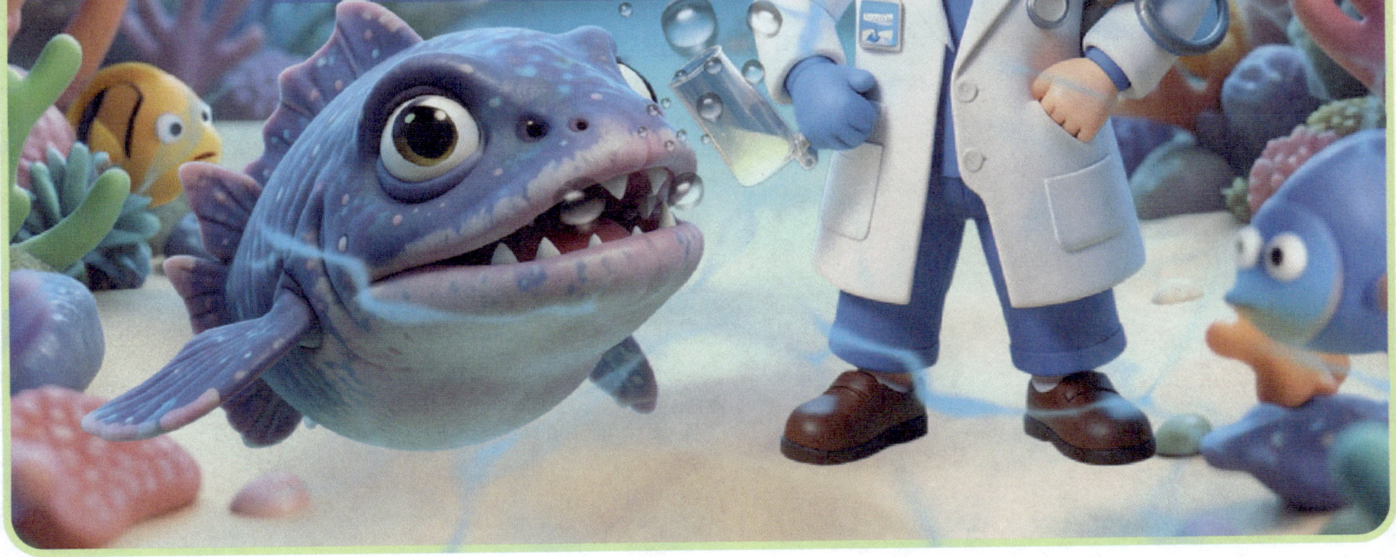

What They Do

Marine biologists study life in the oceans.
That includes:

- Whales, dolphins, and sharks
- Fish, turtles, and coral reefs
- Tiny plankton you can barely imagine
- Strange creatures living deep in the dark sea

They study how ocean animals:

- Eat and move
- Communicate
- Protect themselves
- Survive in changing water conditions

Some marine biologists work underwater.
Others work in labs or with computers.

All of them are trying to understand how ocean life work sand how to protect it.

Where You See Their Work in Real Life

You see marine biologists' work when:

- You visit an aquarium and read the signs
- You watch ocean documentaries
- Fishing rules protect certain species
- Rescue teams help injured whales or turtles

Marine biologists also help explain how climate change affects oceans, coral reefs, and sea animals.
Even though most of us live on land, the oceans affect everyone.

Cool Tools & Tech

- Marine biologists use tools that help them study a world that's hard to reach, such as:
- Underwater cameras and drones
- Diving gear and small submarines
- GPS trackers placed on animals
- Water testing kits
- Computer models that track ocean temperatures and currents

Wow fact:

👉 The oceans cover about 70% of Earth, yet humans have explored less than the surface of the Moon.

Future Problem They Help Solve

Marine biologists help solve problems like:

- Saving coral reefs from warming and bleaching
- Protecting animals from plastic pollution
- Understanding how oceans absorb carbon dioxide
- Keeping seafood safe and sustainable

Without marine biologists, we'd know very little about most of our planet.

Try It Now: Ocean Spotter (Observe)

You'll need:

- Paper
- Pencil

Steps:

- Write down 5 animals you think live in the ocean.
- Circle the ones that breathe air.
- Circle the ones that breathe underwater.
- Ask: How do scientists study animals they can't visit easily?

You'd Like This If You...

- Love ocean animals and nature
- Don't mind getting wet, sandy, or salty
- Care about protecting the planet
- Enjoy learning outdoors or traveling
- Like science that feels adventurous

Why this matters to you:

Healthy oceans help control Earth's climate, provide food, and support life everywhere—even far from the sea.

Marine biologists protect a world most people never see.

And by understanding what lives beneath the waves, they help make sure the ocean stays alive for generations to come.

Geneticist – DNA Code Cracker

Imagine This...

Inside almost every cell in your body is a tiny instruction book. It decides things like:

- Your eye color
- Your hair texture
- How tall might you grow
- How your body fights sickness

You can't see this book.
You can't feel it.
But it's there quietly giving directions every second of your life.
If secrets, codes, and "how did that happen?" questions excite you, you might already be thinking like a **geneticist.**

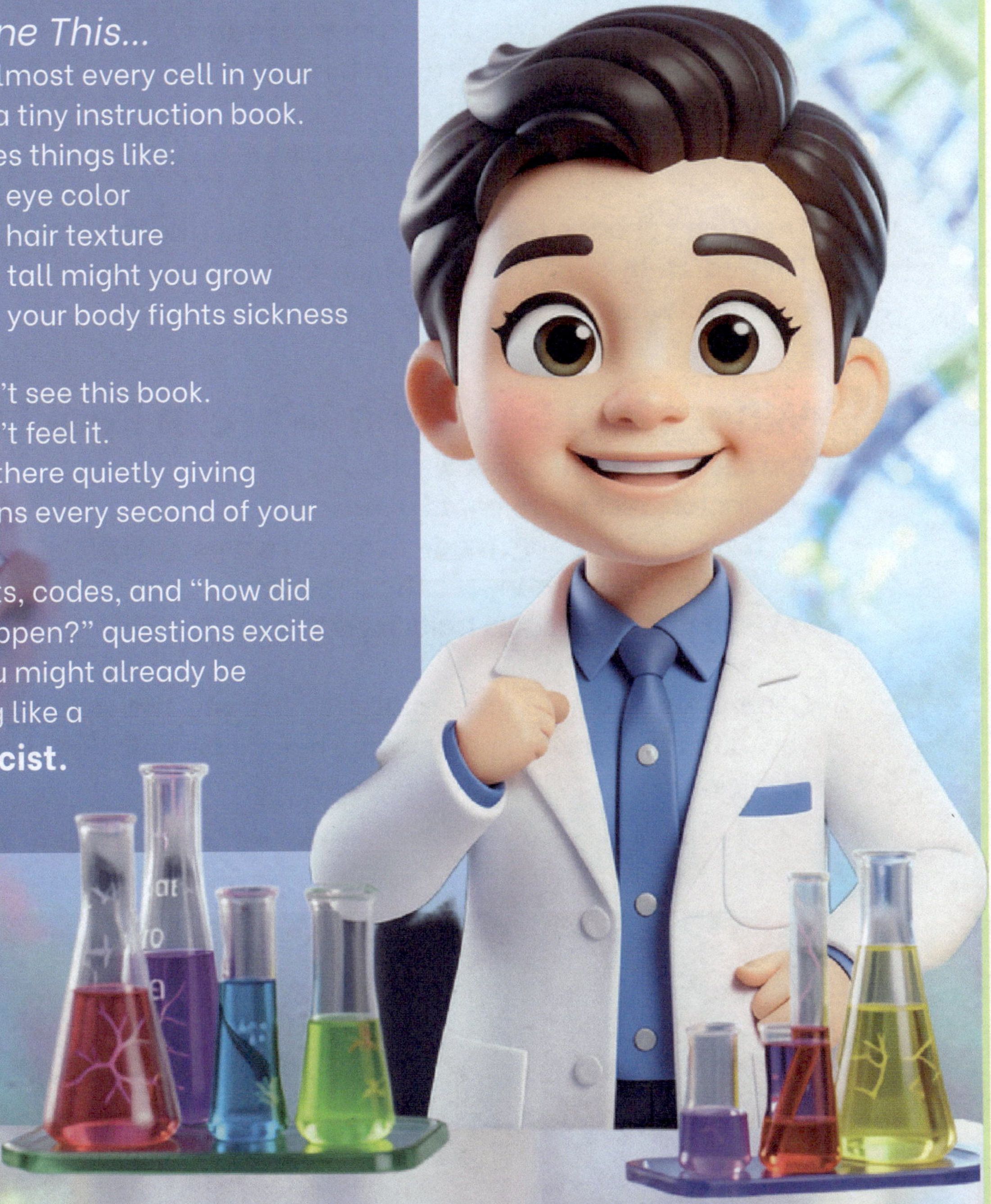

What They Do

Geneticists study DNA.
DNA is like a code made of tiny letters that tells living things how to grow and work.

Geneticists try to understand:

- How traits are passed from parents to kids
- Why do some diseases run in families
- How cells know what to do
- What happens when DNA changes

They don't change people into superheroes (that's movies).
They carefully study how life's instructions work—and how to fix problems when something goes wrong.

Where You See Their Work in Real Life

You see geneticists' work when:

- Doctors test for inherited diseases
- New medicines are made to match a person's DNA
- Farmers grow stronger or healthier crops
- Scientists study how traits pass through families

Geneticists also help solve mysteries, like why two siblings can look very different—or why some people react differently to the same medicine.

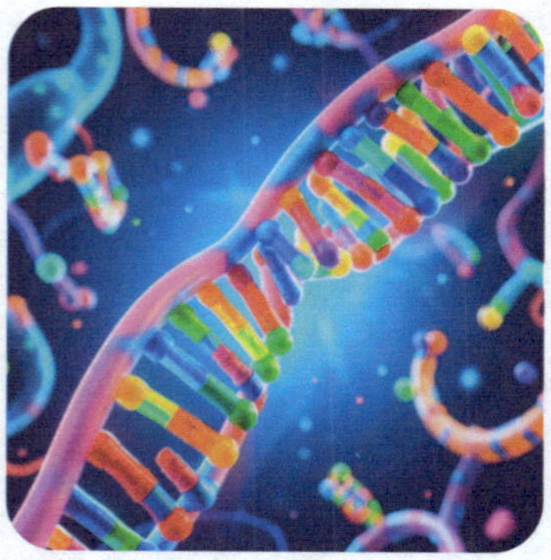

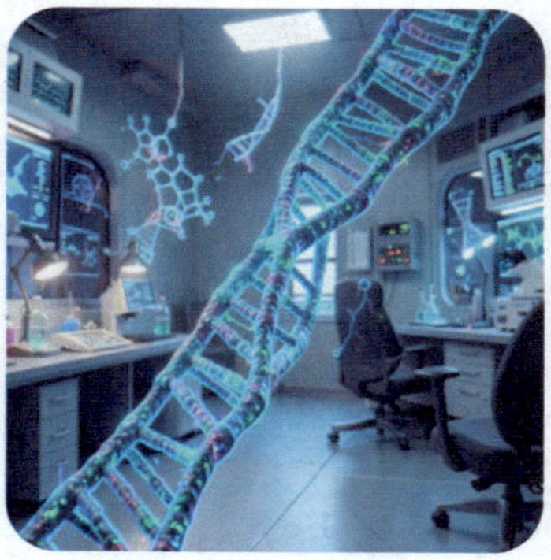

Cool Tools & Tech

Geneticists use tools that help them read and understand DNA, such as:

- DNA sequencers that read the genetic code
- Microscopes to study cells
- Lab machines that copy DNA
- Computers to compare genetic patterns

Wow fact:

👉 If you stretched out all the DNA in just one of your cells, it would be about 6 feet long—and you have trillions of cells.

Future Problem They Help Solve

Geneticists help solve big problems, including:

- Finding treatments for genetic diseases
- Creating personalized medicine that works better for each person
- Helping crops survive droughts or disease
- Understanding how humans evolved

Some geneticists even study ancient DNA to learn about people who lived thousands of years ago.

Try It Now: Ocean Spotter (Observe)

You'll need:

- A mirror
- Family members (optional)

Steps:

- Look at your hair, eyes, and smile.
- Compare with a family member.
- List what's similar and what's different.
- Ask: Which traits might come from genes?

You'll notice patterns—and mysteries.
That's how genetics begins: noticing similarities and differences.

You’d Like This If You...

- Love puzzles and secret codes
- Curious about families and traits
- Enjoy science that explains why things happen
- Like careful, detailed work
- Wonder how tiny changes can make big differences

Why this matters to you:

Understanding DNA helps doctors treat people better and helps scientists solve problems that affect families and future generations.

Geneticists don’t just study the past.

They help shape healthier futures—one tiny code at a time.

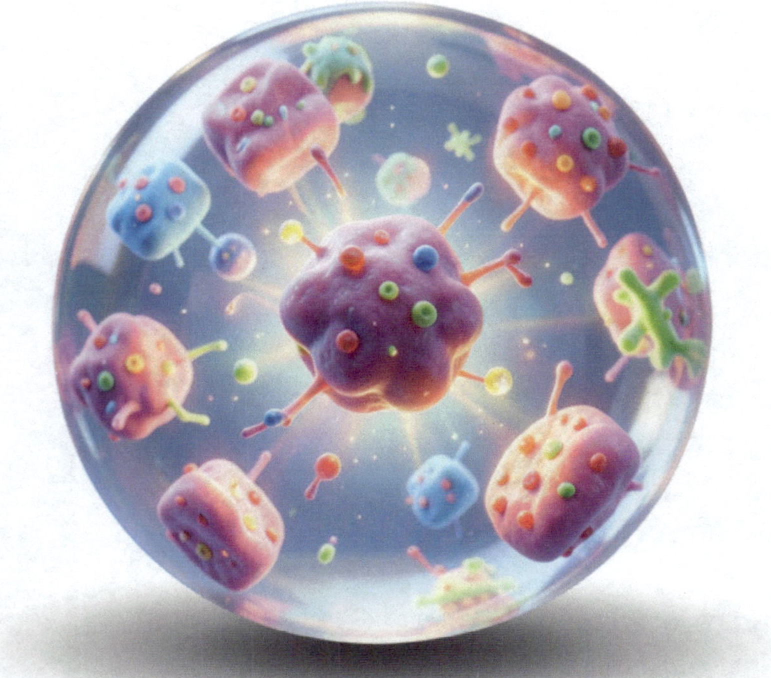

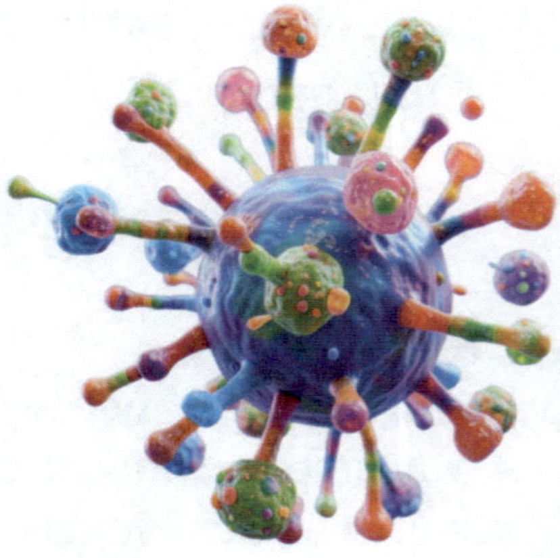

Neuroscientist – Brain Explorer

Imagine This...

You're trying to remember where you left your shoes.
You feel nervous before a test.
You laugh at a joke without even thinking about it.
All of that is happening because of a soft, wrinkly organ inside your head.
Your brain.
It weighs about as much as a small melon—but it controls everything you do.
If you've ever wondered why people think, feel, learn, or behave the way they do, you might already be thinking like a **neuroscientist.**

What They Do

Neuroscientists study the **brain and nervous system.**
They try to understand:

- How we think and remember
- Why do we feel emotions
- How our senses work
- What happens when the brain gets sick or injured

Some neuroscientists study tiny brain cells called neurons. Others study behavior, learning, sleep, or emotions.

All of them are trying to answer one big question:
How does the brain make us who we are?

Where You See Their Work in Real Life

You see neuroscientists' work when:

- Doctors use brain scans to help patients
- New treatments help people with anxiety or depression
- Scientists improve how students learn and remember
- Therapists help people recover after brain injuries

Neuroscience also helps design:

- Brain-training games
- Better learning tools
- Safer technology that works with how brains focus and rests

Every time someone understands their mind a little better, neuroscience is at work.

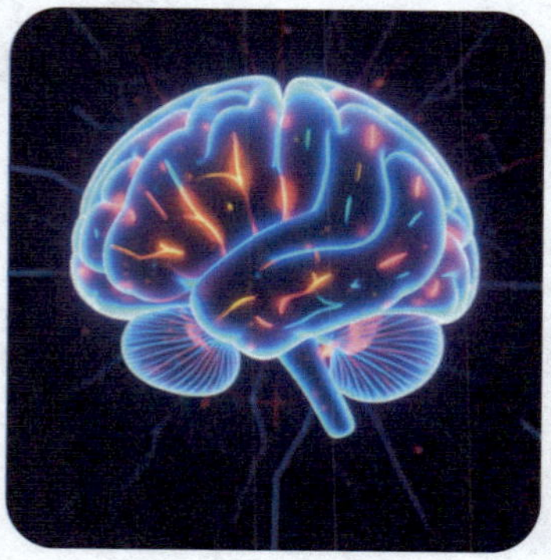

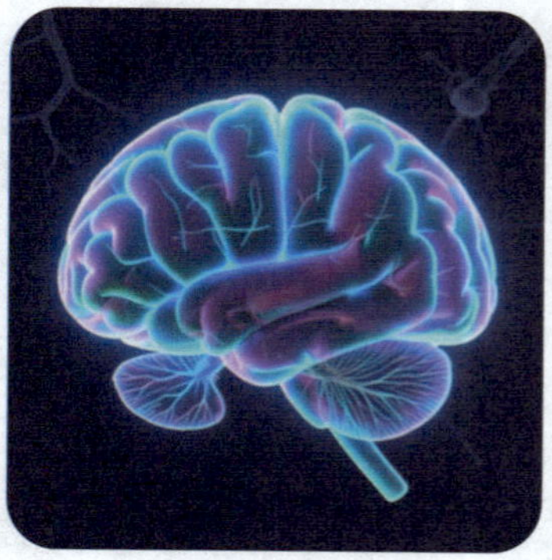

Cool Tools & Tech

Neuroscientists use tools that let them study the brain safely, such as:

- Brain scanners (like MRI machines)
- EEG caps that read brain signals
- Microscopes to study brain cells
- Computer models of brain networks

Wow fact:

👉 Your brain has **about 86 billion neurons**, and each one can connect to thousands of others creating trillions of connections.
That's more connections than there are stars in our galaxy.

Future Problem They Help Solve

Neuroscientists help solve problems like:

- Treating brain diseases like Alzheimer's
- Helping people recover from strokes or injuries
- Understanding mental health better
- Designing schools and technology that match how brains actually learn

Some neuroscientists are even helping people control devices using just their thoughts.

Try It Now: The Color-Word Brain Test

Write these words in big letters:

- RED
- BLUE
- GREEN
- YELLOW

Color each word using a different color than the word says.

Ask someone to:

- First: read the words out loud
- Then: say the color of the ink, not the word

Which was harder?
That confusion shows how your brain handles mixed signals—something neuroscientists study all the time.

You'd Like This If You...

- Love puzzles and secret codes
- Curious about families and traits
- Enjoy science that explains why things happen
- Like careful, detailed work
- Wonder how tiny changes can make big differences

Why this matters to you:

Understanding DNA helps doctors treat people better and helps scientists solve problems that affect families and future generations.

Geneticists don't just study the past.

They help shape healthier futures—one tiny code at a time.

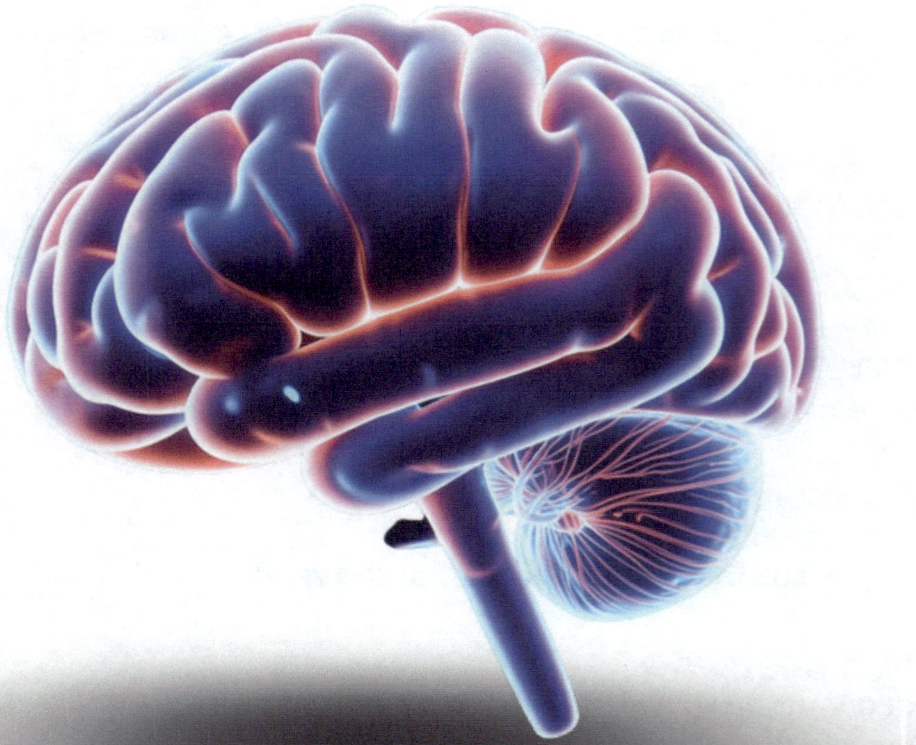

Ecologist – Nature Pattern Finder

Imagine This...

You're outside.
Maybe it's a park, a forest, or even just your backyard.
You notice birds flying in certain places.
You see plants growing better in one spot than another.
You wonder why some areas feel full of life and others don't.
Most people enjoy nature and move on.
You stop and look closer.
If you like noticing patterns in nature and wondering how living things depend on each other, you might already be thinking like an **ecologist.**

What They Do

Ecologists study how living things work together.
They don't just study one animal or one plant.
They study systems.
That means they look at:

- Plants
- Animals
- Soil
- Water
- Weather
- Humans

Ecologists ask questions like:

- Why do certain plants grow here but not there?
- What happens if one species disappears?
- How do animals, plants, and the environment affect each other?

They are pattern finders who try to understand how nature stays balanced—and what happens when that balance changes.

Where You See Their Work in Real Life

You see ecologists' work when:

- Nature parks are protected
- Fishing or hunting rules are created
- Scientists explain why certain animals are endangered
- Cities plan green spaces

Ecologists help decide:

- Where it's safe to build
- How to protect forests and rivers
- How can humans live without harming nature

Their work helps both wildlife and people.

Cool Tools & Tech

Ecologists use tools that help them study nature carefully, such as:

- Field notebooks for observations
- Cameras and drones to watch animals
- GPS trackers to follow movement
- Soil and water testing kits
- Computer models to study ecosystems

Wow fact:

👉 A single tree can support **hundreds of different living organisms**, from insects to birds to tiny microbes.

Future Problem They Help Solve

Ecologists help solve problems like:

- Protecting endangered species
- Preventing ecosystem collapse
- Managing forests, oceans, and wetlands
- Helping nature recover after disasters

As climate changes, ecologists help communities understand how to protect food, water, and wildlife.

Try It Now: Build a Nature Map

You'll need:

- Paper
- Pencil

Steps:

1. Choose a small outdoor area (yard, park, or school grounds).
2. Draw what you see: plants, insects, birds, water, paths.
3. Draw arrows showing connections (who eats what, who uses shade, where water flows).

You just created a simple ecosystem map.
That's exactly how ecologists start their work.

You'd Like This If You...

- Love being outdoors
- Notice patterns in nature
- Care about protecting animals and plants
- Like big-picture thinking
- Want to help humans live in balance with nature

Why this matters to you:

Understanding ecosystems helps protect the places you love and the life that depends on them—including us.

Ecologists don't just study nature.

They help nature survive—and help people learn how to live wisely inside it.

Materials Scientist
Inventor of New Stuff

Imagine This...

Your phone slips from your hand and doesn't shatter.
A jacket keeps you warm but feels light as air.
A bike helmet protects your head without weighing you down.
Those things didn't happen by accident.
Someone figured out what stuff should be made of and how to make it better.
If you like building, testing, and wondering why one material works better than another, you might already be thinking like a **materials scientist.**

What They Do

Materials scientists study what things are made of and how to improve them.

They work with materials like:

- Metals
- Plastics
- Glass
- Fabrics
- Ceramics

They ask questions like:

- Why is this strong but that one breaks?
- How can we make this lighter without making it weaker?
- Can this material bend, stretch, or heal itself?

Materials scientists look at materials up close—sometimes all the way down to atoms—to understand how tiny changes can make a big difference.

Where You See Their Work in Real Life

You see materials scientists' work everywhere.

You see it when:

- Phone screens don't crack as easily
- Sports gear is lighter and safer
- Shoes are bouncy but supportive
- Cars and planes use strong, lightweight parts
- Medical implants work safely inside bodies

Materials scientists help decide what everything around you is made of—even things you never think about.

Cool Tools & Tech

Materials scientists use tools that help them test and improve materials, such as:

- Powerful microscopes that can see tiny structures
- 3D printers to create and test new designs
- Machines that stretch, bend, or crush materials
- Furnaces that heat materials to very high temperatures
- Lasers and sensors that measure strength and change

Wow fact:

👉 Some materials can heal their own cracks, just like skin heals a cut.

Future Problem They Help Solve

Materials scientists help solve problems like:

- Creating eco-friendly materials that don't pollute
- Making batteries last longer and charge faster
- Building lighter vehicles that use less energy
- Designing materials for space travel and extreme environments

They help make technology safer, stronger, and better for the planet.

Try It Now: Material Match (Observe)

You'll need:

- 3 household items (plastic, metal, fabric)

Steps:

1. Touch each item.
2. Ask: Which is hard? Soft? Flexible?
3. Think: Why was each material chosen for its job?

That's exactly the kind of question materials scientists ask every day.

You'd Like This If You...

- Love building and experimenting
- Enjoy science mixed with design
- Curious why some things feel "cheap," and others feel "strong"
- Like testing ideas instead of guessing
- Want to invent better tools, gadgets, or gear

Why this matters to you:

Better materials make life safer, easier, and more sustainable—and they shape the technology you'll use in the future.

Materials scientists don't just study stuff.

They design what the future will be made of.

Forensic Scientist
Real-Life Crime Solver

Imagine This...

A mystery needs answers.
Something happened, but no one saw it clearly.
There are no obvious clues.
Just tiny details left behind.
A fingerprint on glass.
A strand of hair.
A few grains of dust on a shoe.
Most people would miss them.
You wouldn't.
If you like mysteries, careful thinking, and letting facts not guesses tell the story, you might already be thinking like a **forensic scientist.**

What They Do

Forensic scientists use **science to solve mysteries.**
They don't usually chase bad guys or carry handcuffs.
They work in labs, studying clues that others can't understand.
They might:

- Analyze fingerprints
- Study DNA from hair or blood
- Examine fibers, paint, glass, or soil
- Figure out what a substance is made of
- Reconstruct what happened using evidence

Their job is to let the evidence speak.

Where You See Their Work in Real Life

You see forensic science at work when:

- Police identify someone using fingerprints or DNA
- Investigators figure out how a fire starte
- Scientists prove whether the evidence matches a suspect
- Innocent people are cleared using new tests

TV shows make forensic science look fast and flashy.
In real life, it's careful, slow, and extremely detailed.

And that's what makes it powerful.

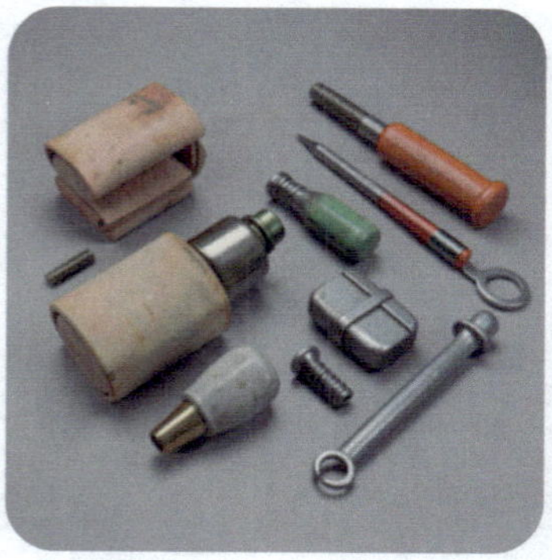

Cool Tools & Tech

Forensic scientists use tools like:

- Microscopes to study tiny details
- DNA sequencers to read genetic material
- Fingerprint scanners
- Chemical testing tools
- Computer databases to compare evidence

Wow fact:

👉 All person's fingerprints are unique—even identical twins don't share the same ones.

Future Problem They Help Solve

Forensic scientists help solve problems like:

- Making investigations more accurate and fairer
- Preventing innocent people from being blamed
- Solving crimes using tiny or damaged evidence
- Investigating digital and cyber crimes

As technology improves, forensic scientists help make sure the truth stays stronger than guesses.

Try It Now: DIY Fingerprint Lab

You'll need:

- A pencil
- Clear tape
- White paper

Steps:

1. Scribble a dark patch with the pencil.
2. Rub your finger on the graphite.
3. Press your finger onto the sticky side of the tape.
4. Place the tape on the paper.
5. Look closely at the lines and swirls.

Now compare fingerprints with a family member.
No two will be exactly the same.

You'd Like This If You...

- Love mystery books and puzzles
- Are careful and patient
- Enjoy science mixed with logic
- Like noticing small details
- Believe facts should matter

Why this matters to you:

Forensic science helps protect truth, fairness, and justice—and makes sure decisions are based on evidence, not assumptions.

Forensic scientists don't guess what happened.

They prove it—one tiny clue at a time.

Food Scientist
Snack & Safety Scientist

Imagine This...

You open a bag of chips, and they're still crunchy.
Milk stays fresh in the fridge.
A frozen meal heats evenly instead of turning half-cold, half-hot.
That didn't happen by luck.
Someone tested, measured, and experimented to make sure food is safe, tasty, and lasts long enough to eat.
If you've ever wondered why food tastes the way it does or how it stays safe, you might already be thinking like a

food scientist.

What They Do

Food scientists study food and how it behaves.
They ask questions like:

- Why does bread rise?
- What makes food spoil?
- How can food stay fresh longer without being unhealthy?
- How can snacks taste good and be safe to eat?

Food scientists mix science with creativity.
They work with chemistry, biology, and sometimes even engineering to understand food from the inside out.

Where You See Their Work in Real Life

You see food scientists' work when:

- Food has expiration dates
- Labels list ingredients clearly
- Snacks taste the same every time you buy them
- Food is packaged to stay fresh
- New foods or flavors appear in stores

They also help make sure food is safe in schools, hospitals, and grocery stores.

Cool Tools & Tech

Food scientists use tools such as:

- Lab equipment to test ingredients
- Temperature sensors to study heating and cooling
- Mixers and test kitchens
- Microscopes to examine food structure
- Computer systems to track food safety

Wow fact:

👉 Many of the foods you eat today were invented in a lab before they ever reached a kitchen.

Future Problem They Help Solve

Food scientists help solve problems like:

- Feeding a growing population
- Reducing food waste
- Creating healthier snacks
- Making food safer during emergencies

They help design food that lasts longer, uses fewer resources, and still tastes good.

Try It Now: Simple Freshness Test

You'll need:

- Two slices of bread
- Two sealable bags
- A warm place

Steps:

1. Put one slice of bread in a bag and seal it tightly.
2. Put the other slice in a bag and leave it slightly open.
3. Place both in the same warm spot.
4. Check them each day for changes.

Which one grows mold first?
Food scientists run tests like this to learn how air and moisture affect food.

You'd Like This If You...

- Like cooking or baking
- Curious about how food works
- Enjoy science, you can taste
- Care about health and safety
- Want to help people eat better

Why this matters to you:

Understanding food science helps you make smarter food choices and keeps people safe and healthy every day.

Food scientists don't just create snacks.

They protect the food that fuels our lives.

Sports Scientist
Body & Motion Coach

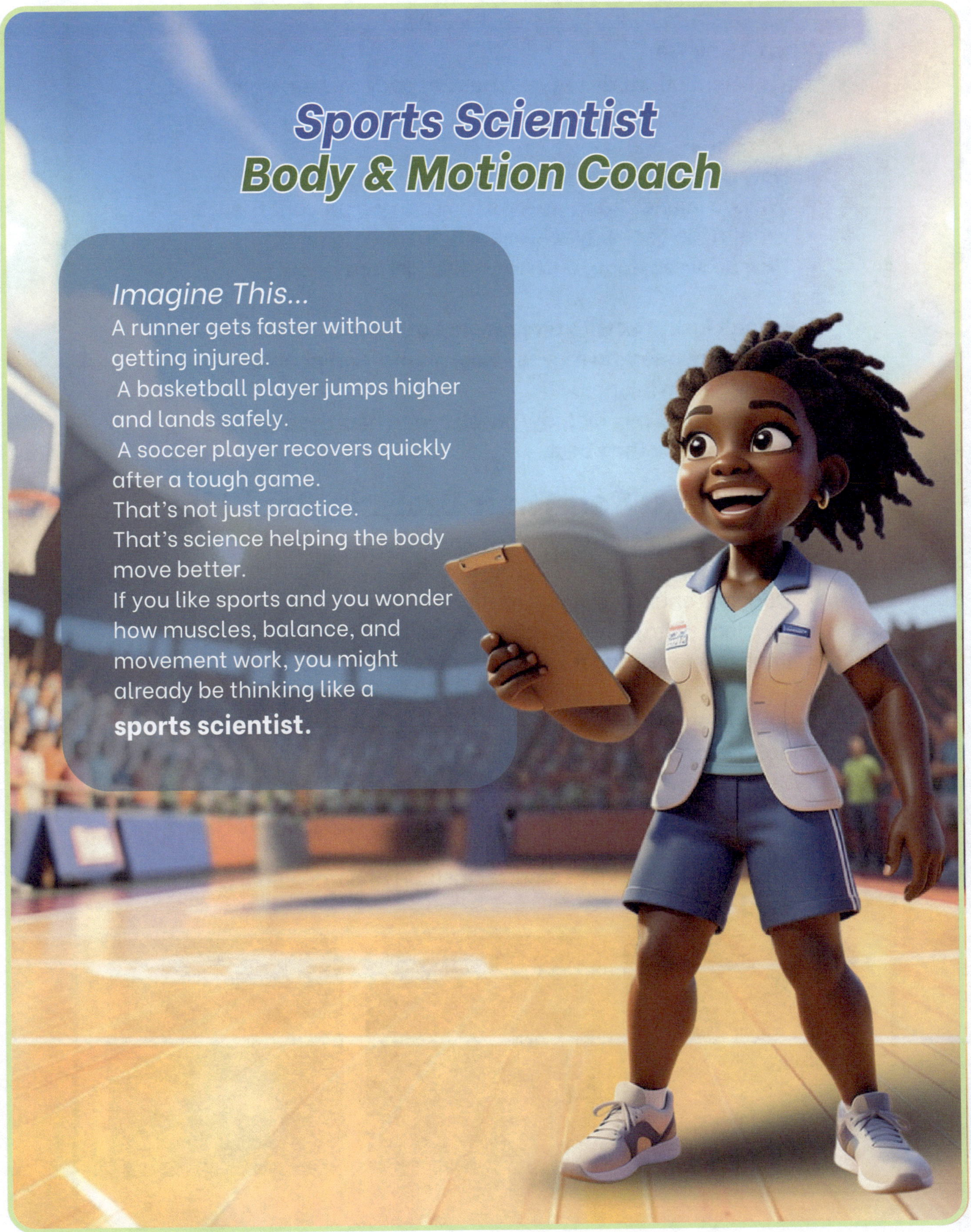

Imagine This...

A runner gets faster without getting injured.
A basketball player jumps higher and lands safely.
A soccer player recovers quickly after a tough game.
That's not just practice.
That's science helping the body move better.
If you like sports and you wonder how muscles, balance, and movement work, you might already be thinking like a **sports scientist.**

What They Do

Sports scientists study how the human body moves, trains, and recovers.

They ask questions like:

- How do muscles grow stronger?
- Why do injuries happen—and how can we prevent them?
- What's the best way to train without overdoing it?
- How do sleep, food, and rest affect performance?

They don't just work with professional athletes.

They help anyone who moves—kids, adults, and people recovering from injuries.

Sports scientists mix biology, physics, and health science to help bodies work their best.

Where You See Their Work in Real Life

You see sports scientists' work when:

- Athletes wear trackers during practice
- Coaches change training routines to avoid injuries
- Physical therapy helps someone walk or run again
- Fitness apps suggest workouts and recovery days

They also help design:

- Safer sports equipment
- Better warm-ups and cool-downs
- Training plans for schools and teams

Anytime someone moves smarter instead of harder, sports science is involved.

Cool Tools & Tech

Sports scientists use tools such as:

- Motion-capture cameras to study movement
- Wearable trackers that measure heart rate and steps
- Force plates that measure jumps and landings
- Stopwatches and sensors
- Computer models of muscles and joints

Wow fact:

👉 When you sprint, some muscles can pull with a force several times your body weight.
Your body is stronger than it looks.

Future Problem They Help Solve

Sports scientists help solve problems like:

- Reducing sports injuries
- Helping people stay active as they age
- Designing safer school sports programs
- Improving recovery after surgery or accidents

They also help people with disabilities move more freely using science-based training and technology.

Try It Now: Your Movement Test

You'll need:

- A stopwatch or timer
- A clear space

Steps:

1. Time how long can you balance on one foot.
2. Switch feet and time again.
3. Now bend your arms or close your eyes slightly and try again.

What changed?
Sports scientists study balance, control, and movement just like this—only with more tools.

You'd Like This If You...

- Love sports or physical activity
- Curious about muscles and movement
- Like science that helps people directly
- Care about staying healthy, not just winning
- Enjoy helping others improve

Why this matters to you:

Understanding how your body moves helps you stay strong, avoid injuries, and enjoy activity for life.

Sports scientists don't just coach athletes.

They help people move better—every step, jump, and stretch of the way.

Climate Scientist
Weather & Future Tracker

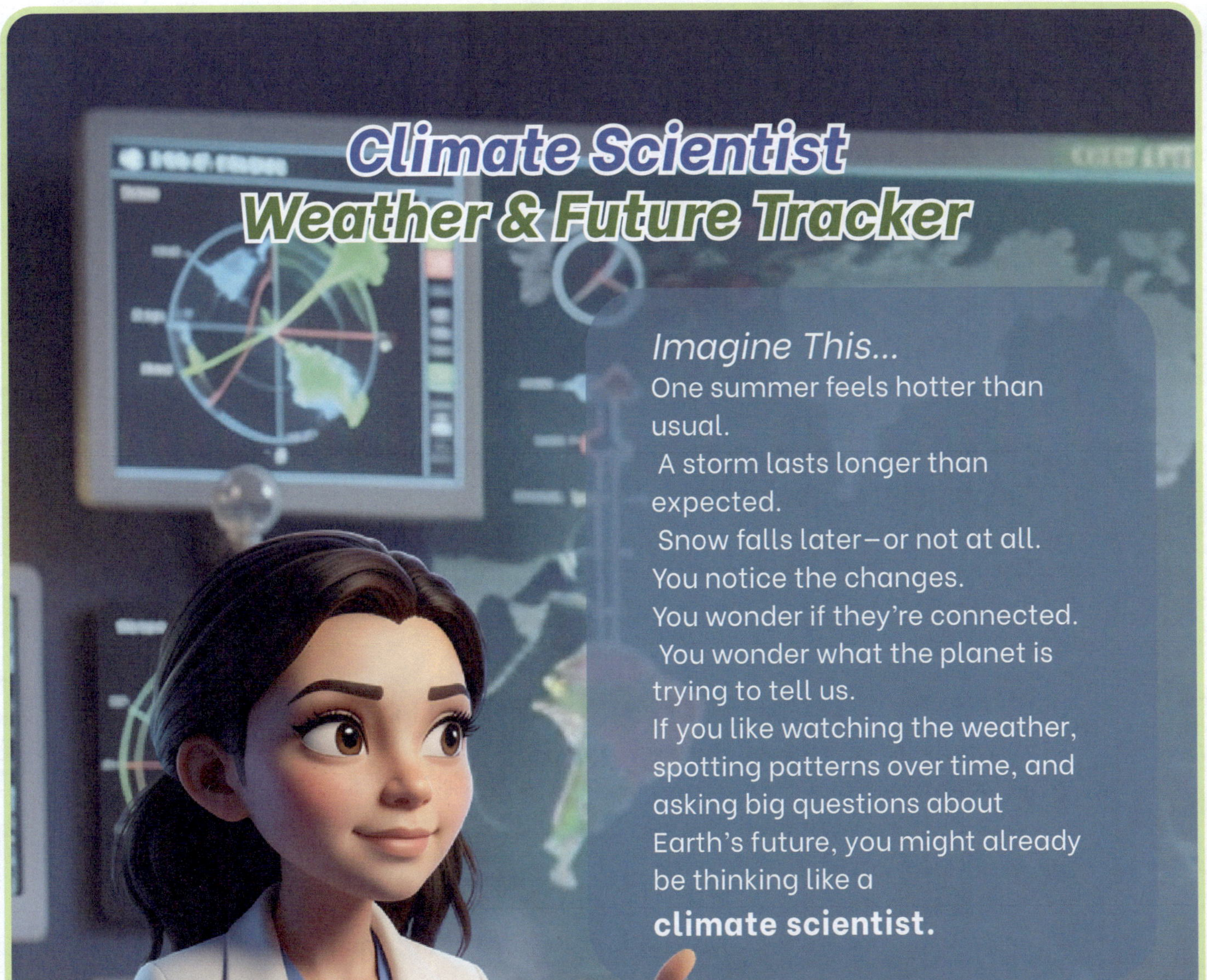

Imagine This...

One summer feels hotter than usual.
A storm lasts longer than expected.
Snow falls later—or not at all.
You notice the changes.
You wonder if they're connected.
You wonder what the planet is trying to tell us.
If you like watching the weather, spotting patterns over time, and asking big questions about Earth's future, you might already be thinking like a

climate scientist.

What They Do (Kid Version)

Climate scientists study **Earth's climate**—the long-term patterns of weather over many years.
They ask questions like:

- Why is Earth warming or cooling?
- How do oceans, air, land, and ice affect each other?
- What happens when climates change too fast?
- How will today's choices affect future life?

Climate scientists don't just look at today's weather. They look at patterns across decades—and sometimes thousands of years.
Their goal is to understand what's changing, why it's happening, and what we can do about it.

Where You See Their Work in Real Life

You see climate scientists' work when:

- Weather apps predict heat waves or storms
- Scientists explain rising sea levels
- Cities plan for floods or droughts
- Schools talk about climate change
- News reports show melting ice or stronger storms

Climate science helps communities prepare, adapt, and protect people and nature.

Cool Tools & Tech

Climate scientists use tools that help them study the whole planet, such as:

- Satellites that watch Earth from space
- Weather stations that collect data
- Ocean sensors that measure temperature and currents
- Ice core samples that reveal ancient climates
- Computer models that simulate Earth's future

Wow fact:

👉 Air bubbles trapped in ancient ice can tell scientists what Earth's atmosphere was like hundreds of thousands of years ago.

Future Problem They Help Solve

Climate scientists help solve problems like:

- Predicting extreme weather events
- Protecting cities from floods and heat
- Helping farmers grow food in changing climates
- Protecting animals and ecosystems
- Guiding governments toward smarter decisions

Their work helps people plan for the future instead of being surprised by it.

Try It Now: Your Movement Test

You'll need:

- A notebook
- A calendar

Steps:

1. Pick one weather feature to track (temperature, rain, clouds).
2. Write it down once a day for two weeks.
3. Look for patterns.

Did things change slowly or quickly?
That's how climate science begins: watching patterns over time.

You'd Like This If You...

- Like weather and nature
- Notice patterns over days and weeks
- Care about the planet's future
- Enjoy science that connects many systems
- Want to help people prepare, not panic

Why this matters to you:

Understanding climate helps protect homes, food, water, and wildlife—and helps people make better choices for the future.

Climate scientists don't control the weather.

They help us understand Earth's story—and where it may be heading next.

PART 2

Tech Builders

Turning Ideas Into Tools People Can Use

Technology isn't just about screens.
It's about **solving problems with tools.**

Tech builders are the people who look at a problem and think, "There has to be a better way to do this."

They design apps, games, systems, robots, and digital spaces that help people learn, connect, stay safe, and get things done. Some work with code. Some work with design. Some work behind the scenes, making sure everything runs smoothly.

Tech builders don't need to know everything.

They experiment.
They test.
They fix mistakes.
They try again.

As you read these careers, notice what excites you:

- Making things work
- Making things easier
- Making things safer
- Making things more fun

Technology is always changing. That means there's room for new ideas and new thinkers.

Maybe one of them is you.
Let's meet the people who build the digital world you use every day.

Software Engineer
App & Game Maker

An app freezes right when you need it.
A game crashes, and all your progress is gone.
Someone has to figure out what broke—and how to fix it.
But behind every tap, click, and swipe is a long list of instructions someone wrote on purpose.
If you like games, puzzles, and figuring out how things work behind the scenes, you might already be thinking like a **software engineer.**

What They Do

Software engineers write instructions for computers.
Those instructions are called code.
Computers don't guess.
They don't "sort of" understand.
They do exactly what they're told.

Software engineers decide:

- What the computer should do first
- What should happen next
- What to do if something goes wrong

They build:

- Apps
- Games
- Websites
- Programs that help doctors, teachers, and scientists

It's like teaching a very fast, very literal robot how to think—step by step.

Where You See Their Work in Real Life

You see software engineers' work when:

- You play a game on a phone or console
- You use a school website or learning app
- You send messages or video chat
- You stream videos or music
- A map app shows you the fastest route

Even traffic lights, elevators, and washing machines use software. If it runs on a screen—or has a tiny computer inside—software engineers helped build it.

Cool Tools & Tech

Software engineers use tools such as:

- Programming languages (like Scratch, Python, or JavaScript)
- Computers and tablets
- Code editors where they write and fix instructions
- Debugging tools to find mistakes
- Version control systems to track changes

Wow fact:

👉 One small mistake in code—like a missing symbol—can stop an entire program from working.
That's why software engineers test everything.

Future Problem They Help Solve

Software engineers help solve problems like:

- Making technology safer and more secure
- Creating learning tools for schools
- Building apps that help people stay healthy
- Designing games that teach teamwork and creativity
- Helping scientists analyze huge amounts of data

As technology grows, software engineers help decide how helpful—or harmful—it becomes.

Try It Now: Be a Human Computer

You'll need:

- Paper
- Pencil
- A friend or family member

Steps:

1. Write instructions for your "human computer" to draw a simple shape (like a square).
2. Be very specific: where to start, how far to move, when to turn.
3. Let them follow your instructions exactly—no guessing.

Did it work the first time?
That's what coding feels like: testing, fixing, and trying again.

You'd Like This If You...

- Love games or apps
- Enjoy logic and puzzles
- Like solving problems step by step
- Don't mind fixing mistakes
- Want to build things other people can use

Why this matters to you:

Understanding software helps you create tools instead of just using them—and gives you power to shape how technology affects the world.

Software engineers don't just play games.

They build the worlds inside them.

AI / Machine Learning Engineer
Teaches Computers to Think

A computer can recognize your face—but it doesn't actually see. It can answer questions—but it doesn't really think.

So how does a machine learn anything at all?

If you've ever wondered how machines seem to get "smarter" over time, you might already be thinking like an

AI or machine learning engineer.

What They Do

AI and machine learning engineers teach computers how to learn from examples.

Instead of giving a computer every rule, they:

- Show it lots of data
- Help it notice patterns
- Let it improve through practice

It's a bit like teaching a puppy.
You don't explain everything with words.
You show examples, correct mistakes, and try again.

AI engineers build systems that can:

- Recognize faces or voices
- Translate languages
- Recommend videos or games
- Help doctors spot problems
- Help cars notice roads and signs

The computer doesn't think like a human, but it can learn patterns very quickly.

Where You See Their Work in Real Life

You see AI engineers' work when:

- A phone unlocks using your face
- A video app suggests what to watch next
- A voice assistant understands your question
- Games adjust difficulty as you play
- Maps predict traffic

AI is also used in schools, hospitals, science labs, and even space missions.

Cool Tools & Tech

AI engineers use tools like:

- Large sets of data (pictures, words, numbers)
- Programming languages
- Training models that learn from examples
- Computers with strong processing power
- Testing tools to check fairness and accuracy

Wow fact:

👉 Some AI systems can learn from millions of examples in just hours something that would take humans years.

Future Problem They Help Solve

AI engineers help solve problems like:

- Detecting diseases earlier
- Making learning more personalized
- Helping robots assist people safely
- Reducing waste and energy use
- Making technology fairer and less biased

They also think carefully about ethics—making sure AI is used to help people, not harm them.

Try It Now: Be a Human Computer

You'll need:

- Paper
- Pencil

Steps:

1. Think about how you learned to recognize animals, letters, or faces.
2. Write down three examples that helped you learn (pictures, repetition, mistakes).
3. Ask: What would a computer need to learn the same thing?

This is how AI learning begins—examples first, rules later.

You'd Like This If You...

- Like patterns and logic
- Are curious about how machines learn
- Enjoy math mixed with creativity
- Care about using technology responsibly
- Like improving systems over time

Why this matters to you:

Understanding AI helps you shape technology instead of letting it shape you and helps make sure future tools are used wisely.

AI engineers don't just build smart machines.

They decide how smart tools help the world.

Cybersecurity Specialist
Internet Bodyguard

Someone clicks a link they shouldn't.
A password is too easy to guess.
Suddenly, private information isn't private anymore.

If you like secrets, spotting weaknesses, and keeping people safe without being seen, you might already be thinking like a **cybersecurity specialist.**

What They Do

Cybersecurity specialists protect computers, networks, and information.
They work to stop:

- Hackers
- Viruses
- Scams
- Data theft

They think in two directions at once:

- How a system is supposed to work
- How someone might try to break it

That means they often think like the bad guys—but use that knowledge for good.
Their job is to find weak spots before someone else does.

Where You See Their Work in Real Life

You see cybersecurity specialists' work when:

- Websites ask for strong passwords
- Apps send security alerts
- Schools protect student information
- Banks keep money safe online
- Games block cheating or fake accounts

Hospitals, power stations, and even traffic systems rely on cybersecurity to stay safe.
If it's connected to the internet, someone needs to protect it.

Cool Tools & Tech

Cybersecurity specialists use tools like:

- Encryption to lock information
- Firewalls to block unwanted access
- Security scanners to find weaknesses
- Password managers
- Monitoring systems that watch for strange activity

Wow fact:

👉 Some hackers can break into systems in seconds if security is weak which is why cybersecurity specialists are always testing and improving defenses.

Future Problem They Help Solve

Cybersecurity specialists help solve problems like:

- Protecting hospitals and schools from attacks
- Keeping personal data private
- Stopping scams and online bullying
- Securing self-driving cars and smart homes
- Making the internet safer for kids and families

As the world connects more devices, cybersecurity becomes more important than ever.

Try It Now: Build a Strong Password System

You'll need:

- Paper
- Pencil

Steps:

1. Think of a short sentence you can remember.
2. Take the first letter of each word.
3. Add numbers or symbols in places you'll remember.

Example:
"I like pizza on Fridays" → Ilp0F!
Now try to guess a friend's password.
Was it easy or hard?
Cybersecurity is about making things hard to guess but easy to remember.

You'd Like This If You...

- Like secrets and puzzles
- Enjoy thinking ahead
- Care about protecting people
- Like logic and problem-solving
- Want to use tech powers for good

Why this matters to you:

Understanding cybersecurity helps you stay safer online—and helps protect others from harm in the digital world.

Cybersecurity specialists don't seek attention.

They quietly stand guard—keeping the internet safer, one system at a time.

Robotics Engineer
Robot Creator

What if a machine could help clean your room?
What if it could explore places too dangerous for people?
What would it need to move, sense, and decide?
Robots don't appear by magic. Someone designed them to move, sense, and act.
If you like building things, making them move, and mixing code with real-world objects, you might already be thinking like a
robotics engineer.

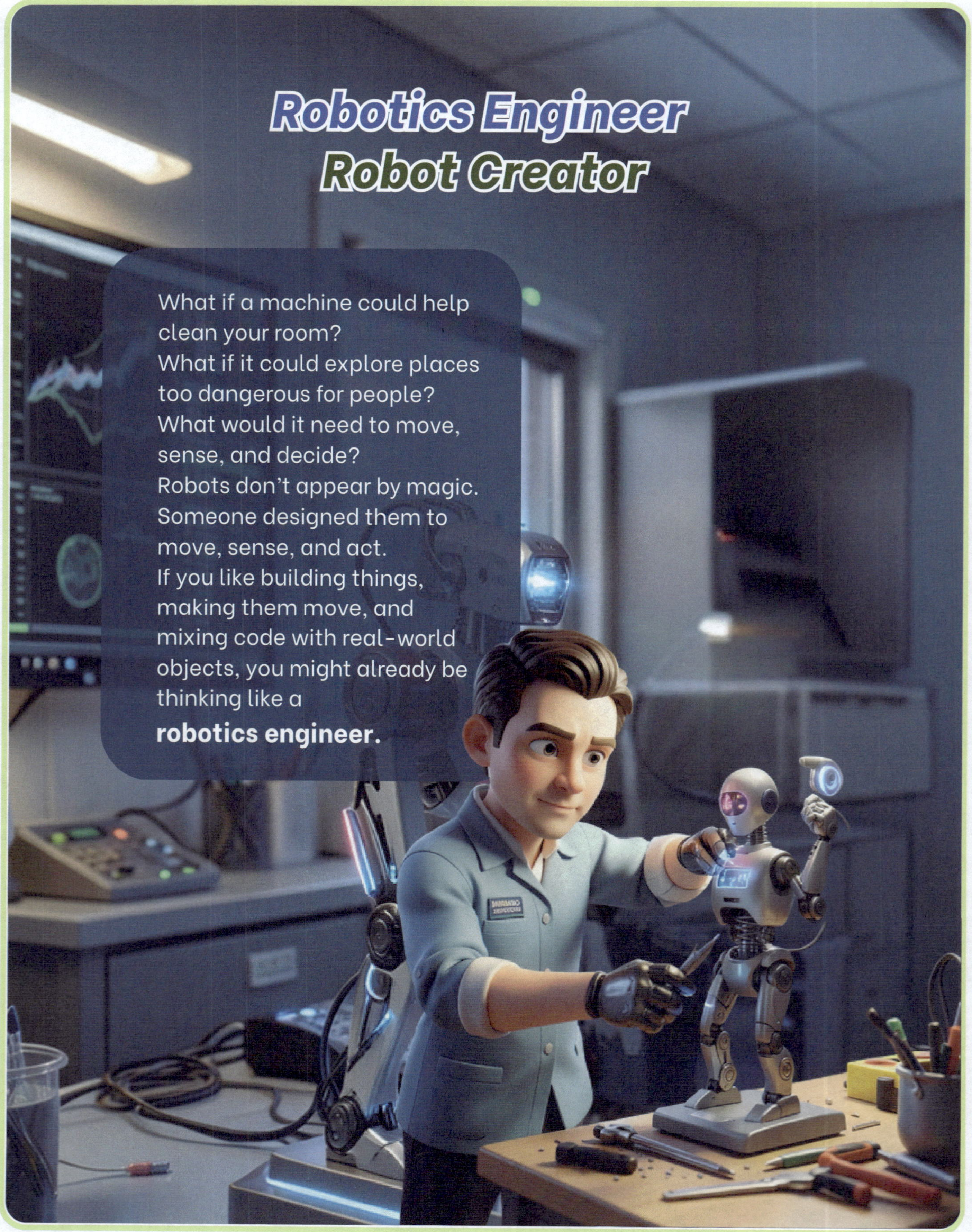

What They Do

Robotics engineers design and build robots that can do tasks.
They combine:

- Engineering (how parts fit and move)
- Programming (what the robot should do)
- Sensors (how the robot understands its surroundings)

Robots can:

- Pick up objects
- Follow lines or maps
- Avoid obstacles
- Help people in dangerous or boring jobs

Robotics engineers don't just build robots.
They teach machines how to interact with the real world.

Where You See Their Work in Real Life

You see robotics engineers' work when:

- Factories use robot arms
- Hospitals use surgical robots
- Warehouses move packages automatically
- Space agencies send robots to Mars
- Homes use robot vacuums

Robots are also used in farming, disaster rescue, and research.

Cool Tools & Tech

Robotics engineers use tools like:

- Motors and gears to create movement
- Sensors to detect light, sound, and distance
- Microcontrollers (tiny computers)
- Programming software
- 3D printers to build parts

Wow fact:

👉 Some robots can react **faster than a human blink.**

Future Problem They Help Solve

Robotics engineers help solve problems like:

- Helping doctors perform safer surgeries
- Exploring dangerous environments
- Assisting people with disabilities
- Improving farming and food production
- Reducing boring or risky jobs

As robots become more common, engineers focus on making them helpful, safe, and trustworthy.

Try It Now: Design a Robot Helper

You'll need:

- Paper
- Pencil

Steps:

1. Think of a problem a robot could help with.
2. Draw a robot that could solve it.
3. Label its parts: wheels, arms, sensors, controller.

That's how many real robots begin—as drawings.

You'd Like This If You...

- Love building and tinkering
- Enjoy code and machines together
- Like solving physical problems
- Don't mind testing and fixing mistakes
- Want to create helpful technology

Why this matters to you:

Robots can help people in ways humans can't—and understanding how they work lets you shape that future.

Robotics engineers don't just imagine the future.

They build it—piece by piece.

UX Designer
Makes Apps Easy (and Fun)

Some apps feel easy the very first time you use them. Others feel confused right away.
That feeling doesn't happen by accident.
Someone carefully designed how the app feels to use.
If you like helping people, noticing small details, and thinking about how things could be easier or more fun, you might already be thinking like a **UX designer.**

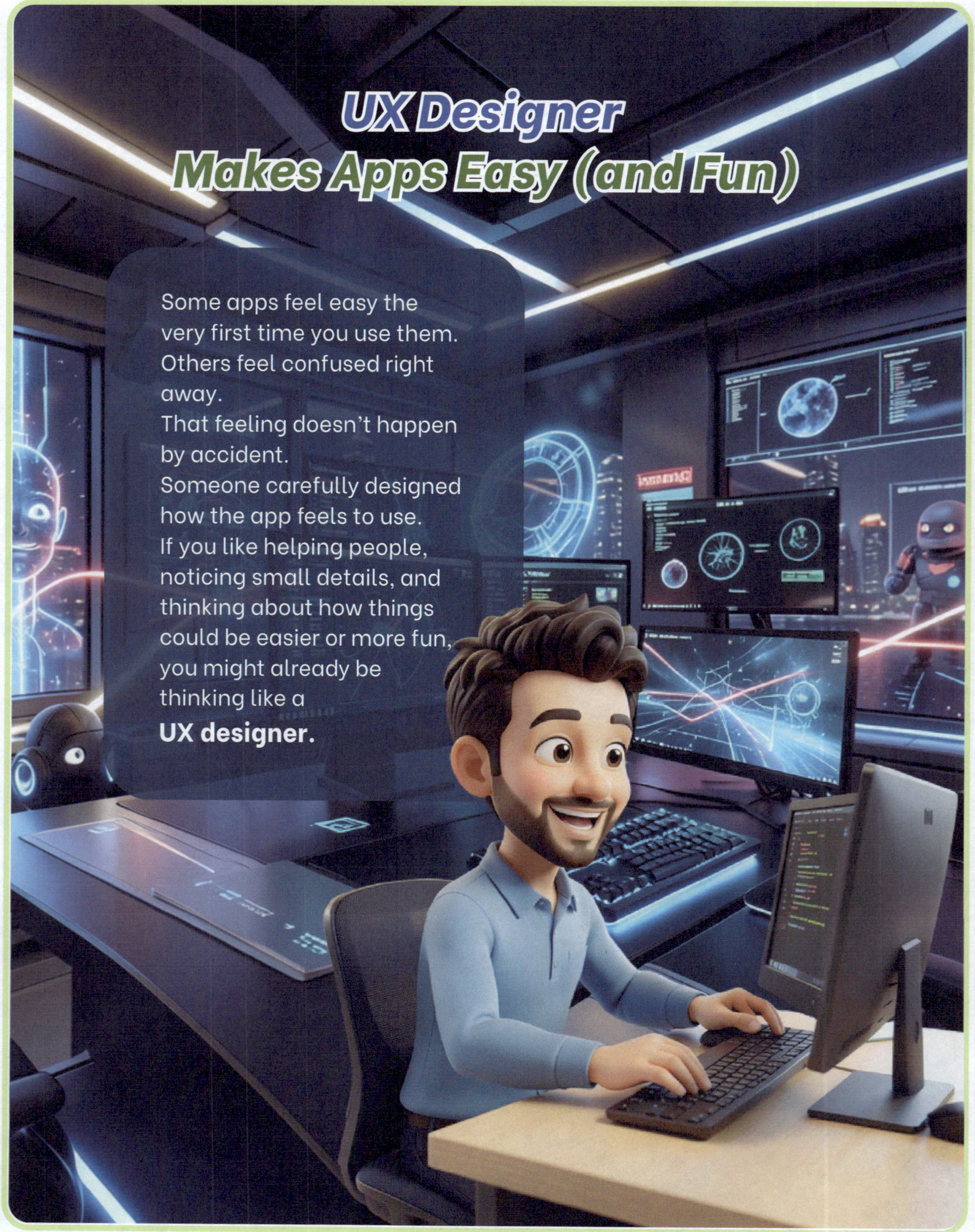

What They Do

UX designers focus on how people experience technology.
"UX" means User Experience.
They ask questions like:

- Is this easy to understand?
- Where do people get stuck or confused?
- Does this feel fun or frustrating?
- Can everyone use this—even kids or older adults?

UX designers don't just care if something works.
They care if it works well for humans.
They design:

- App layouts
- Button placement
- Menus and screens
- How steps flow from one to the next

Their goal is to make technology feel friendly, not stressful.

Where You See Their Work in Real Life

You see UX designers' work when:

- Apps are easy to use without instructions
- Websites load clearly on phones and tablets
- Games guide you smoothly from level to level
- Forms are simple instead of confusing
- Accessibility features help more people use tech

If you've ever thought, "This app is so easy," a UX designer helped make that happen.

Cool Tools & Tech

UX designers use tools like:

- Sketches and wireframes (simple drawings of screens)
- Prototypes you can click through
- User testing sessions to watch how people use an app
- Design software
- Feedback notes from real users

Wow fact:

👉 UX designers often watch people use apps without giving help, just to see where they struggle.
Watching mistakes helps make designs better.

Future Problem They Help Solve

UX designers help solve problems like:

- Making technology accessible for everyone
- Reducing frustration and screen stress
- Designing safer apps for kids
- Helping people use important tools quickly in emergencies
- Making learning apps more engaging

As technology grows, UX designers help make sure it stays human friendly.

Try It Now: Design Your Own App Screen

You'll need:

- Paper
- Pencil or markers

Steps:

1. Think of an app you use often.
2. Draw one screen you think could be better.
3. Decide where buttons should go and why.
4. Show your design to someone and see if they understand it.

That's exactly how UX designers test ideas.

You'd Like This If You...

- Care about how people feel
- Like art mixed with problem-solving
- Notice when things are confusing
- Enjoy helping others
- Want to make technology kinder and easier

Why this matters to you:

Good design saves time, reduces stress, and helps more people use technology confidently.

UX designers don't just make things look nice.

They make technology feel right.

Data Scientist
Pattern Detective

One number by itself doesn't mean much.
But many numbers together can tell a powerful story.
The challenge is knowing what the story really says.
Most people see a mess.
You see clues.
If you like sorting, comparing, and asking, "What is this really telling us?" you might already be thinking like a **data scientist.**

What They Do

Data scientists study data–information collected from the world.
That data can come from:

- Sports games
- Weather reports
- Apps and websites
- Health trackers
- Surveys and tests

Data scientists look for:

- Patterns
- Trends
- Surprises
- Answers hidden inside numbers

They don't just collect information.
They turn it into stories that help people make better decisions.

Where You See Their Work in Real Life

You see data scientists' work when:

- Apps suggest what you might like
- Sports teams improve their strategies
- Weather forecasts get more accurate
- Schools study learning progress
- Health experts track disease spread

Anytime numbers help guide a decision, a data scientist is involved.

Cool Tools & Tech

Data scientists use tools like:

- Spreadsheets and databases
- Graphs and charts
- Programming languages
- Data visualization tools
- Computers that handle large datasets

Wow fact:

👉 One website can collect millions of data points every minute. Without data scientists, that information would be useless.

Future Problem They Help Solve

Data scientists help solve problems like:

- Predicting disease outbreaks
- Improving traffic flow
- Reducing waste and pollution
- Making schools and hospitals more effective
- Helping businesses and governments plan wisely

They help turn "too much information" into smart choices.

Try It Now: Design Your Own App Screen

You'll need:

- Paper
- Pencil

Steps:

1. Write down your bedtime for the last three nights.
2. Look for a pattern (earlier, later, the same).
3. Ask: What could this pattern help explain?

👉 Data scientists look for meaning in patterns, not just numbers.

You'd Like This If You...

- Enjoy numbers and patterns
- Like organizing information
- Are curious about "why" something happens
- Enjoy problem-solving
- Want facts to guide decisions

Why this matters to you:

Understanding data helps you make smarter choices—and helps others do the same.

Data scientists don't just crunch numbers.

They find the meaning hiding inside them.

Game Designer
Worlds You Can Play In

A game is so easy that players get bored.
A game is so hard that players quit.
Getting the balance wrong can ruin the fun.
And if you've ever wondered why a game is fun, fair, exciting, or frustrating—and thought about how you would make it better—you might already be thinking like a **game designer.**

What They Do

Game designers create the ideas behind games.
They don't usually draw every picture or write every line of code.
Instead, they decide:

- What is the goal of the game is
- What players can and can't do
- How hard or easy challenges should be
- How the game teaches you to play
- What makes the game fun instead of boring

Game designers mix:

- Creativity
- Storytelling
- Rules
- Psychology

Their job is to design experiences that players want to come back to again and again.

Where You See Their Work in Real Life

You see game designers' work when:

- A level slowly gets harder as you improve
- A tutorial teaches you without long instructions
- A game feels fair, even when you lose
- A story makes you care about the characters
- Multiplayer games balance teamwork and competition

Board games, card games, sports games, and video games all have game designers behind them.

Cool Tools & Tech

Game designers use tools like:

- Game design documents (plans for how games work)
- Paper prototypes (testing ideas without screens)
- Storyboards and flowcharts
- Game engines
- Playtesting sessions where people try the game

Wow fact:

👉 Many famous games started as simple paper sketches before becoming digital worlds.

Future Problem They Help Solve

Game designers help solve problems like:

- Making learning more engaging through games
- Teaching teamwork and problem-solving
- Creating games that are fair and inclusive
- Helping people practice skills safely (like simulations)

Games aren't just for fun—they can teach, train, and connect people.

Try It Now: Design a Game on Paper

You'll need:

- Paper
- Pencil

Steps:

1. Decide your game's goal.
2. Write 3 rules players must follow.
3. Add one challenge that makes it interesting.
4. Test it with a friend and watch what happens.

Did they have fun?
Where did they get confused?
That's exactly how game designers improve their games.

You'd Like This If You...

- Love games and puzzles
- Enjoy storytelling and creativity
- Care about fairness and fun
- Like thinking about how people feel
- Want to build worlds others can explore

Why this matters to you:

Games can teach skills, bring people together, and help us practice solving problems.

Game designers don't just make games.

They design experiences—and invite others to play inside them.

AR / VR Developer Builds Virtual Worlds

What if learning felt like stepping inside a lesson? What if you could visit places without leaving your room?

Technology is starting to make that possible.

And If you like imagining new worlds and using technology to bring them to life, you might already be thinking like an

AR / VR developer.

What They Do

AR and VR developers create digital worlds you can experience, not just look at.

- VR (Virtual Reality) places you fully inside a digital world.
- AR (Augmented Reality) adds digital objects to the real world around you.

They decide:

- What you see
- What you hear
- How you move
- How you interact

They mix:

- Coding
- Design
- 3D art
- Storytelling

Their goal is to make digital experiences feel believable, useful, and exciting.

Where You See Their Work in Real Life

You see AR and VR developers' work when:

- Games place characters into your room
- Virtual tours let you visit museums or landmarks
- Training simulations help pilots or doctors practice safely
- Educational apps let you explore space or history
- Filters add effects to photos and videos

AR and VR aren't just for games—they help people learn, train, and explore.

Cool Tools & Tech

AR / VR developers use tools like:

- VR headsets and motion controllers
- Cameras and sensors to track movement
- 3D modeling software
- Game engines
- Programming languages

Wow fact:

👉 Your brain can react to virtual experiences almost the same way it reacts to real ones–including fear, excitement, and curiosity.

Future Problem They Help Solve

AR and VR developers help solve problems like:

- Training people for dangerous jobs safely
- Helping doctors practice surgeries
- Making education more immersive
- Allowing people to explore places they can't physically visit
- Helping people with anxiety practice real-life situations

As technology improves, virtual worlds will become more helpful–and more realistic.

Try It Now: Design a Virtual World

You'll need:

- Paper
- Pencil

Steps:

1. Choose a place you'd like to explore virtually.
2. Decide what you would see, hear, and interact with.
3. Draw or write your ideas.

That's how many virtual worlds begin–inside someone's imagination.

You'd Like This If You...

- Love imagination and creativity
- Enjoy games and technology
- Like storytelling and design
- Are curious about how brains experience reality
- Want to build new ways to explore the world

Why this matters to you:

Virtual worlds can help people learn, practice, and connect in powerful ways.

AR and VR developers don't just build technology.

They build experiences—and invite others to step inside.

Cloud Engineer
Keeps the Internet's "Sky" Working

You save a photo, and it doesn't disappear.
You start a video on one device and finish it on another.
A game remembers your progress, even after you log out.
It feels like your stuff is floating somewhere… waiting for you.
That "somewhere" is called the cloud.
If you like systems that work quietly in the background and enjoy making sure everything runs smoothly, you might already be thinking like a **cloud engineer.**

What They Do

Cloud engineers build and take care of the internet's storage and power systems.

The cloud is not a real cloud in the sky.

It's made of huge computers called servers that store data and run apps.

Cloud engineers decide:

- Where information is stored
- How fast it can be reached
- How to keep it safe
- What happens if something breaks

They make sure millions—even billions—of people can use apps and websites at the same time without problems.

Where You See Their Work in Real Life

You see cloud engineers' work when:

- Your photos back up automatically
- School files save online
- Games sync across devices
- Videos stream without stopping
- Websites don't crash during busy times

If you've ever thought, "It just works," a cloud engineer helped make that happen.

Cool Tools & Tech

Cloud engineers use tools like:

- Data centers (buildings full of servers)
- Virtual machines that run many systems at once
- Security tools to protect data
- Monitoring dashboards to watch performance
- Automation tools that fix problems fast

Wow fact:

👉 A single data center can use as much electricity as a small town—so cloud engineers also work to make systems more energy-efficient.

Future Problem They Help Solve

Cloud engineers help solve problems like:

- Keeping the internet fast as more people go online
- Protecting data from loss or attacks
- Supporting online learning and remote work
- Reducing energy use in massive computer systems
- Making technology available anywhere in the world

As the internet grows, cloud engineers keep it reliable and ready.

Try It Now: Be a System Thinker

You'll need:

- Paper
- Pencil

Steps:

1. Draw how a photo gets from your phone to the cloud.
2. Add steps like "upload," "store," and "download."
3. Think about what could go wrong at each step—and how to fix it.

That's exactly how cloud engineers think: planning for problems before they happen.

You'd Like This If You...

- Like organizing systems
- Enjoy solving behind-the-scenes problems
- Care about reliability and safety
- Prefer quiet teamwork over attention
- Want to make technology dependable

Why this matters to you:

Reliable cloud systems keep your work, memories, and learning safe and accessible—anytime, anywhere.

Cloud engineers don't get much attention.

But without them, the internet would fall out of the sky.

Bioinformatician
Biology + Coding Explorer

A scientist has millions of pieces of information about cells, genes, or diseases. Way too much for one person to read.

But what if a computer could help?

If you like biology and enjoy patterns, logic, or coding, you might already be thinking like a **bioinformatician.**

What They Do

Bioinformaticians work at the meeting point of biology and computers. They help scientists understand huge amounts of biological information, like:

- DNA sequences
- Protein structures
- How cells behave
- How diseases spread

Instead of looking through data by hand, bioinformaticians write programs that:

- Organize information
- Spot patterns
- Compare changes
- Find answers faster

They turn confusing data into clear insights that scientists can actually use.

Where You See Their Work in Real Life

You see bioinformaticians' work when:

- Doctors study genetic test results
- Scientists track new viruses
- Researchers develop new medicines
- Labs compare DNA samples
- Health experts predict disease spread

Bioinformaticians played a huge role in understanding viruses and helping doctors respond quickly during health emergencies.

Cool Tools & Tech

Bioinformaticians use tools like:

- Computers and servers for large data sets
- Programming languages
- Databases full of biological information
- Visualization tools to map genes and proteins
- Algorithms that compare patterns

Wow fact:

👉 Reading one human genome creates so much data that it would fill thousands of books—which is why computers are essential.

Future Problem They Help Solve

Bioinformaticians help solve problems like:

- Creating personalized medicine
- Finding treatments faster
- Predicting disease outbreaks
- Understanding how life evolves
- Improving food and agriculture

As science collects more data, bioinformaticians make sure it actually helps people.

Try It Now: Pattern Spotter

You'll need:

- Paper
- Pencil

Steps:

1. Write a short pattern using letters (like A, T, C, G).
2. Repeat it with small changes.
3. Circle what stays the same and what changes.

That's how bioinformaticians begin spotting genetic patterns.

You'd Like This If You...

- Enjoy biology and technology
- Like patterns and puzzles
- Are patients with complex information
- Want to help science move faster
- Enjoy working behind the scenes

Why this matters to you:

Bioinformatics helps turn scientific discoveries into real-world solutions that save time—and lives.

Bioinformaticians don't just study life.

They help decode it.

PART 3

Everyday Engineers

Building the World We Live In

Look around you.

The roads you travel on.
The buildings you enter.
The lights that turn on.

The water that flows when you open a tap.

None of it happened by accident.

Everyday, engineers are the people who design, build, and improve the systems that make daily life possible. Their work isn't always flashy, but it's everywhere—and it matters every single day.

They think about safety.
They think about strength.
They think about efficiency.
They think about what could go wrong—and how to prevent it.

As you explore these careers, notice what draws you in:

- Big structures
- Moving machines
- Power and electricity
- Medical tools
- Systems that must work reliably

Everyday, engineers don't just imagine ideas.
They turn ideas into things people depend on.
Let's meet the builders behind the world around you.

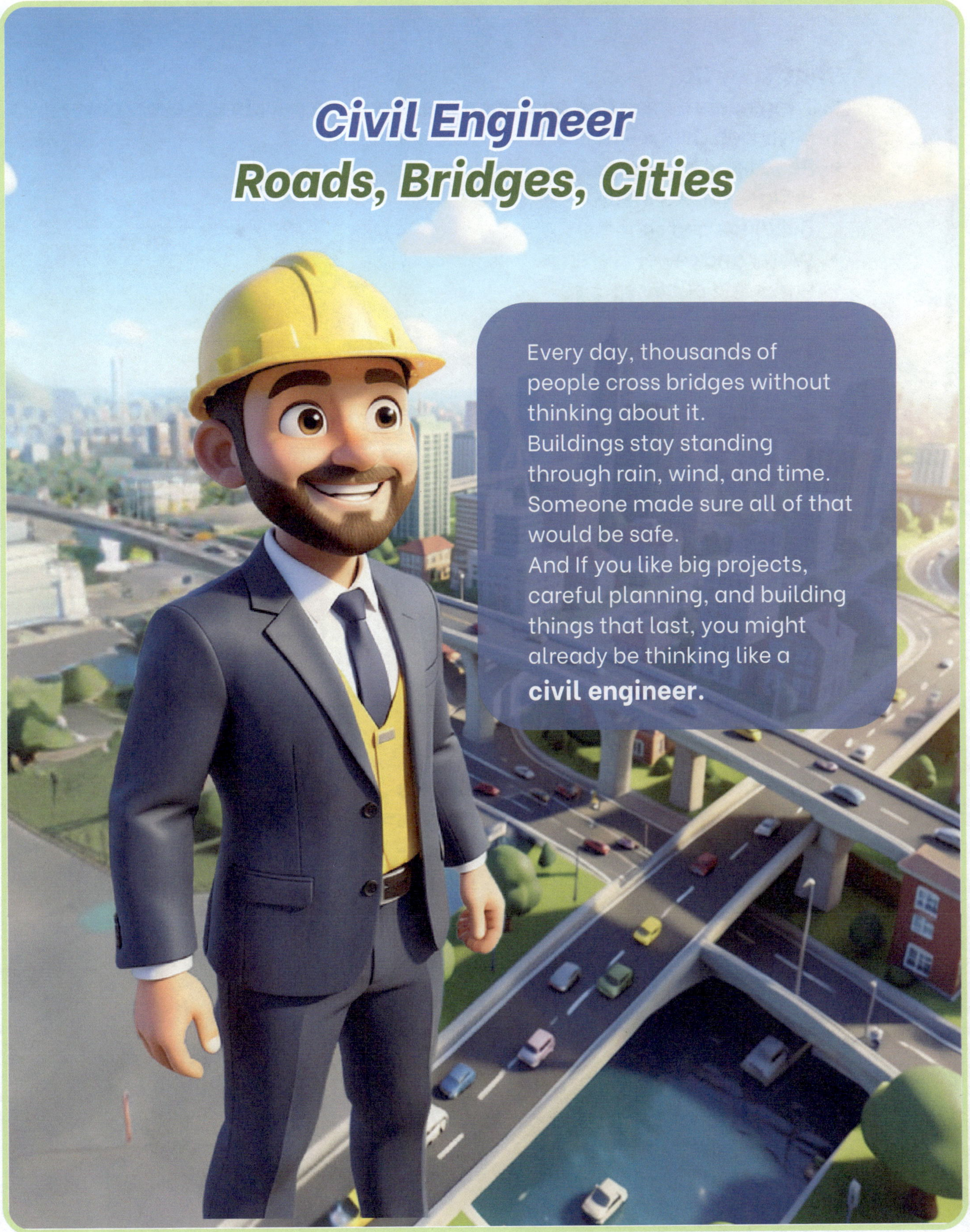

Civil Engineer
Roads, Bridges, Cities

Every day, thousands of people cross bridges without thinking about it.
Buildings stay standing through rain, wind, and time.
Someone made sure all of that would be safe.
And If you like big projects, careful planning, and building things that last, you might already be thinking like a **civil engineer.**

What They Do

Civil engineers design and help build the structures people use every day. They work on things like:

- Roads and highways
- Bridges and tunnels
- Buildings and schools
- Water and sewer systems
- Parks and public spaces

Civil engineers ask questions like:

- Will this be strong enough?
- Is it safe for everyone?
- How long will it last?
- How can we build it better for the future?

Their job is to make sure the places we live and travel are safe, useful, and dependable.

Where You See Their Work in Real Life

You see civil engineers' work when:

- You cross a bridge
- You walk on a sidewalk
- You drink water from a tap
- You enter a school, a mall, or a hospital
- A city adds bike lanes or drainage systems

If a structure supports people every day, a civil engineer helped plan it.

Cool Tools & Tech

Civil engineers use tools such as:

- Blueprints and digital plans
- Computer models that test strength
- Surveying tools to measure land
- Construction materials like concrete and steel
- Safety testing equipment

Wow fact:

👉 Some bridges are designed to bend slightly instead of breaking during strong winds or earthquakes.

Future Problem They Help Solve

Civil engineers help solve problems like:

- Building cities that can handle floods or earthquakes
- Designing safer roads to reduce accidents
- Creating clean water systems for growing populations
- Making cities more accessible for everyone

As cities grow and climates change, civil engineers help make sure communities stay safe.

Try It Now: Spot the Support (Observe)

You'll need:

- Your surroundings
- Paper (optional)

Steps:

1. Look around and find three things that support weight (a chair, stairs, a table, a bridge picture).
2. Ask: What might happen if that support were weaker?
3. Ask: Why is strength important for things people use every day?

👉 Civil engineers start by noticing what must stay strong and safe.

You'd Like This If You...

- Enjoy building and planning
- Like working on big projects
- Care about safety and fairness
- Want to improve cities and communities
- Enjoy solving real-world problems

Why this matters to you:

Civil engineering helps create safe places to live, learn, and travel—places people rely on every day.

Civil engineers don't just build structures.

They build the foundation of everyday life.

Mechanical Engineer
Moving Machines Master

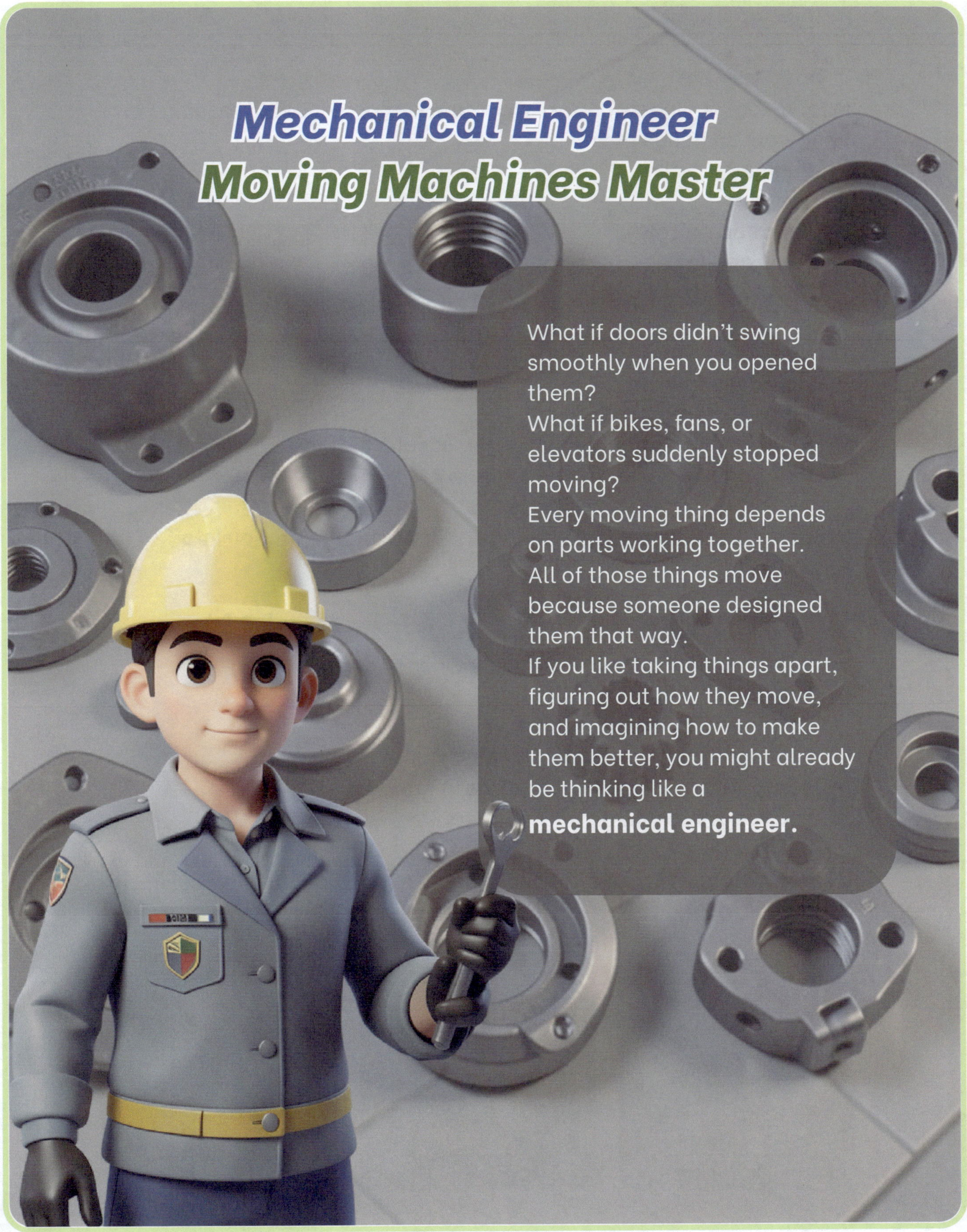

What if doors didn't swing smoothly when you opened them?
What if bikes, fans, or elevators suddenly stopped moving?
Every moving thing depends on parts working together.
All of those things move because someone designed them that way.
If you like taking things apart, figuring out how they move, and imagining how to make them better, you might already be thinking like a **mechanical engineer.**

What They Do

Mechanical engineers design and improve machines with moving parts. They work on things like:

- Engines
- Vehicles
- Fans and motors
- Tools and toys
- Robots and machines

They ask questions like:

- How does this move?
- What parts work together?
- How can it be smoother, stronger, or safer?
- What happens if something breaks?

Mechanical engineers use science and math to turn motion into something useful.

Where You See Their Work in Real Life

You see mechanical engineers' work when:

- You ride a bike or a scooter
- You open a door with hinges
- You use a fan or an air conditioner
- You watch machines build things in factories
- You see amusement park rides in action

If something spins, slides, lifts, or rotates, a mechanical engineer likely helped design it.

Cool Tools & Tech

Mechanical engineers use tools like:

- Gears, levers, and pulleys
- Motors and bearings
- 3D design software
- Prototypes and test machines
- Sensors to measure force and speed

Wow fact:

👉 Tiny machines with moving parts can be smaller than a grain of sand.

Future Problem They Help Solve

Mechanical engineers help solve problems like:

- Designing energy-efficient machines
- Creating safer vehicles and tools
- Building better prosthetic limbs
- Making factories cleaner and quieter
- Improving everyday products

They help turn ideas into machines that work in the real world.

Try It Now: Motion Watcher (Observe)

You'll need:

- Any object with moving parts (door, toy, fan, zipper)

Steps:

1. Watch the object move.
2. Count how many parts move together.
3. Ask: What would happen if one part stopped working?

👉 Mechanical engineers think about motion, forces, and teamwork between parts.

.

You'd Like This If You...

- Love building and tinkering
- Enjoy seeing how things move
- Like hands-on problem-solving
- Don't mind testing and fixing
- Want to design machines that help people

Why this matters to you:

Mechanical engineering helps create tools and machines that make life easier, safer, and more fun.

Mechanical engineers don't just make things move.

They make the world work.

Electrical Engineer
Power & Circuits Pro

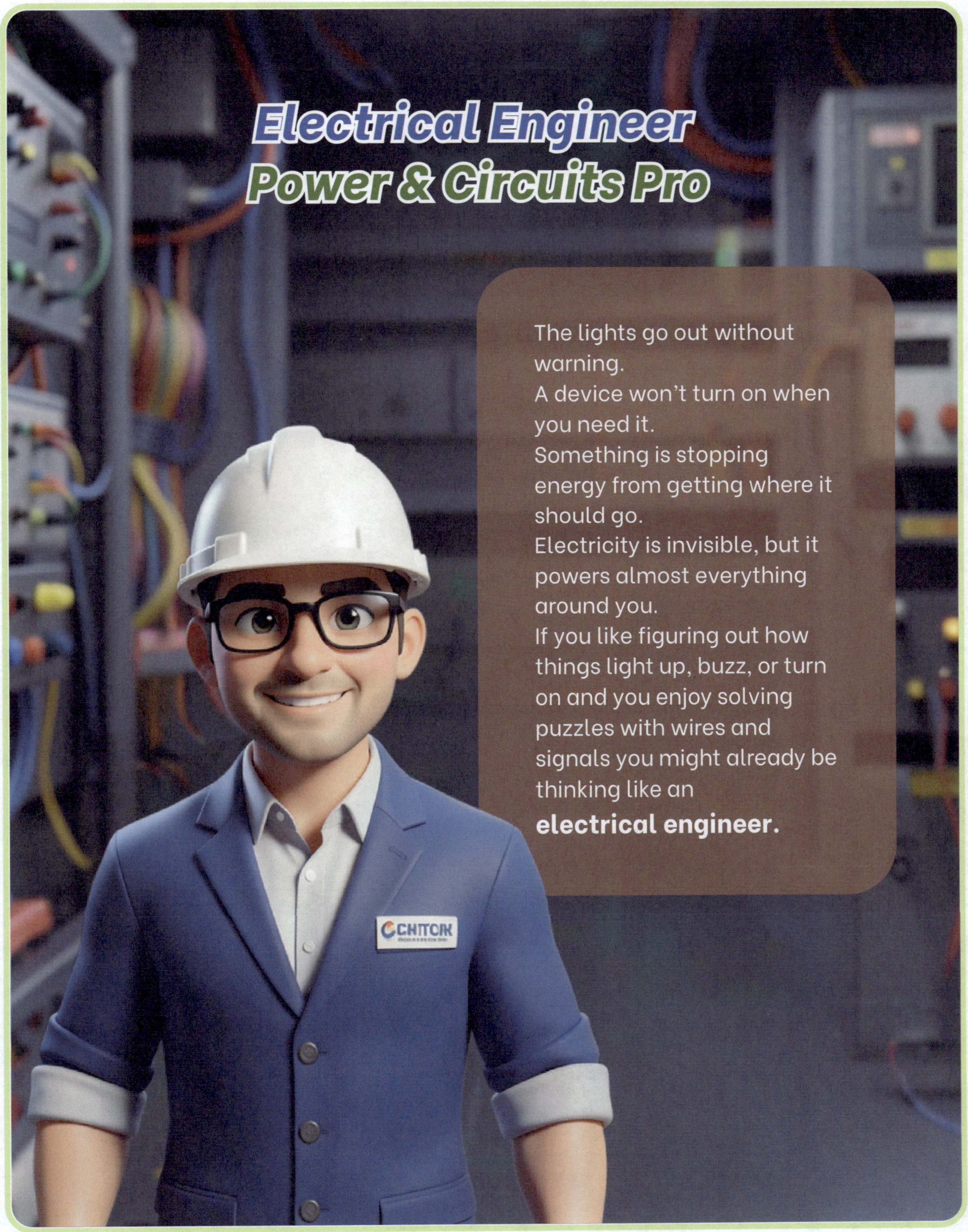

The lights go out without warning.
A device won't turn on when you need it.
Something is stopping energy from getting where it should go.
Electricity is invisible, but it powers almost everything around you.
If you like figuring out how things light up, buzz, or turn on and you enjoy solving puzzles with wires and signals you might already be thinking like an **electrical engineer.**

What They Do

Electrical engineers work with electricity and circuits.
They design systems that:

- Create power
- Control power
- Move power safely

They work on things like:

- Power grids and batteries
- Electronics and circuit boards
- Motors and machines
- Communication systems
- Renewable energy systems

Electrical engineers ask questions like:

- How does electricity flow here?
- How can we control it safely?
- What happens if something overloads?

Their job is to make sure energy gets where it needs to go—without causing danger.

Where You See Their Work in Real Life

You see electrical engineers' work when:

- The lights turn on instantly
- Devices charge safely
- Traffic lights work properly
- Trains and subways run
- Solar panels power homes

From tiny gadgets to entire cities, electrical engineers keep things running.

Cool Tools & Tech

Electrical engineers use tools like:

- Wires, switches, and resistors
- Circuit boards
- Multimeters to measure electricity
- Design software for circuits
- Safety equipment to prevent shocks

Wow fact:

👉 Electricity travels close to the speed of light, which is why lights turn on almost instantly when you flip a switch.

Future Problem They Help Solve

Electrical engineers help solve problems like:

- Making clean energy more reliable
- Designing safer batteries
- Powering electric vehicles
- Bringing electricity to remote areas
- Reducing energy waste

As the world uses more electricity, electrical engineers help make it safer and smarter.

Try It Now: Energy Path Thinker (Observe)

You'll need:

- Your surroundings
- Paper (optional)

Steps:

1. Look at something that uses electricity (light, fan, charger).
2. Trace where the energy comes from and where it goes.
3. Ask: What happens if the path is broken?

👉 Electrical engineers think about how energy flows—and how to keep it safe.

You'd Like This If You...

- Enjoy puzzles and logic
- Like building with wires or electronics
- Are curious about how power works
- Care about safety and precision
- Want to power the future

Why this matters to you:

Understanding electricity helps you use technology safely—and helps bring power to people who need it most.

Electrical engineers don't just work with energy.

They keep the world switched on.

Biomedical Engineer Builds Medical Tech

A doctor uses a machine to see inside the human body without surgery.
A person walks again using a robotic leg.
A heart keeps beating because of a tiny device inside someone's chest.
Those life-changing tools didn't appear by accident.
If you like science and want to help people feel better, stronger, or healthier, you might already be thinking like a

biomedical engineer.

What They Do

Biomedical engineers design technology that works with the human body.

They mix:

- Biology (how the body works)
- Engineering (how to build things)
- Medicine (how to help people heal)

They create and improve things like:

- Artificial limbs and braces
- Heart monitors and pacemakers
- Medical imaging machines
- Surgical tools
- Devices that help people move, breathe, or hear

Their goal is simple but powerful:
Use technology to help people live better lives.

Where You See Their Work in Real Life

You see biomedical engineers' work when:

- Doctors use machines to check your heart
- X-rays or scans show what's happening inside the body
- People use hearing aids or insulin pumps
- Hospitals use safer, smarter equipment
- Athletes recover faster with medical support

Anytime technology helps doctors understand or treat the body, biomedical engineers are involved.

Cool Tools & Tech

Biomedical engineers use tools such as:

- Medical sensors that measure heart rate or oxygen
- 3D printers to create custom implants
- Computer models of bones, muscles, and organs
- Lab equipment to test materials safely
- Design software for medical devices

Wow fact:

👉 Some medical implants are designed to last 20 years or more inside the human body.

Future Problem They Help Solve

Biomedical engineers help solve problems like:

- Helping people recover from injuries
- Creating better treatments for diseases
- Making medical care safer and faster
- Designing affordable medical devices for everyone
- Improving life for people with disabilities

As medicine advances, biomedical engineers help turn discoveries into real tools doctors can use.

Try It Now: Design for Comfort (Reflect)

You'll need:

- Paper
- Pencil

Steps:

1. Think of something people wear or use for health (glasses, braces, bandages).
2. Write one thing that could feel uncomfortable.
3. Ask: How could the design be changed to help the person more?

👉 Biomedical engineers think carefully about how devices feel and function in real bodies.

You'd Like This If You...

- Care deeply about helping others
- Enjoy science and engineering together
- Like solving meaningful problems
- Are curious about how the body works
- Want to make a real difference

Why this matters to you:

Biomedical engineering helps people heal, move, and live more fully—often in ways they never thought possible.

Biomedical engineers don't just build machines.

They build hope.

Aerospace Engineer
Planes & Rockets

Planes are made of metal
but they stay in the sky.
Rockets push upward
instead of falling back down.
Flying seems impossible...
yet it works.
If you love airplanes, space,
and asking "How does that
stay up?", you might already
be thinking like an
aerospace engineer.

What They Do

Aerospace engineers design vehicles that fly.

They work on:

- Airplanes
- Helicopters
- Rockets
- Satellites
- Spacecraft

They think about:

- How air flows around wings
- How much force is needed to lift off
- How materials handle heat and pressure
- How to keep pilots and astronauts safe

Aerospace engineers mix physics, math, and creativity to make flight possible.

Where You See Their Work in Real Life

You see aerospace engineers' work when:

- Planes take off and land smoothly
- Satellites help GPS and weather forecasts
- Rockets launch into space
- Drones fly steadily
- Space missions send back photos

Anytime something flies, aerospace engineers help make it happen.

Cool Tools & Tech

Aerospace engineers use tools like:

- Wind tunnels to test airflow
- Flight simulators
- Computer models of aircraft
- Strong, lightweight materials
- Heat shields for spacecraft

Wow fact:

👉 A rocket must travel faster than 25,000 kilometers per hour to stay in orbit around Earth.

Future Problem They Help Solve

Aerospace engineers help solve problems like:

- Making planes use less fuel
- Designing quieter aircraft
- Sending humans safely to space
- Exploring other planets
- Improving satellites that help life on Earth

They help connect Earth to the skies—and beyond.

Try It Now: Why Does It Fly? (Observe)

You'll need:

- A sheet of paper

Steps:

1. Drop the paper straight down and watch how it falls.
2. Now gently throw it forward.
3. Ask: What changed when the paper moved through the air?

👉 Aerospace engineers study how air and motion work together.

You'd Like This If You...

- Love flight and space
- Enjoy physics and problem-solving
- Like precision and careful testing
- Dream big but think carefully
- Want to explore beyond Earth

Why this matters to you:

Aerospace engineering helps humans travel, communicate, and explore in ways once thought impossible.

Aerospace engineers don't just look up.

They figure out how to get there.

Environmental Engineer
Cleaner Air & Water

A river becomes unsafe to drink from.
The air in a city makes people sick.
Someone has to figure out how to fix the damage humans create.
And if you care about the planet and like fixing problems humans have caused, you might already be thinking like an **environmental engineer.**

What They Do

Environmental engineers design systems that protect nature and human health.

They work on problems like:

- Cleaning dirty water
- Reducing air pollution
- Managing waste and recycling
- Protecting soil and land
- Preventing environmental disasters

They mix science, engineering, and creativity to figure out how people and nature can live together safely.

Environmental engineers don't just study problems.

They build solutions.

Where You See Their Work in Real Life

You see environmental engineers' work when:

- Water from the tap is safe to drink
- Recycling systems work properly
- Factories reduce harmful waste
- Cities improve air quality
- Lakes and rivers are cleaned and protected

Anytime you see nature being repaired or protected with technology, environmental engineers are involved.

Cool Tools & Tech

Environmental engineers use tools like:

- Water filters and treatment systems
- Air quality sensors
- Computer models of pollution movement
- Lab equipment to test samples
- Renewable energy systems

Wow fact:

👉 Some filters can clean thousands of liters of water a day—using only gravity.

Future Problem They Help Solve

Environmental engineers help solve problems like:

- Providing clean water to everyone
- Reducing plastic and chemical pollution
- Protecting communities from floods
- Making cities greener and healthier
- Fighting climate-related damage

As the planet changes, environmental engineers help people adapt without destroying nature.

Try It Now: Why Does It Fly? (Observe)

You'll need:

- A plastic bottle (cut in half)
- Sand, gravel, and cloth
- Dirty water (muddy, not dangerous)

Steps:

1. Layer cloth, sand, and gravel in the bottle.
2. Pour the dirty water through.
3. Observe how the water looks afterward.

Real filters are more advanced—but the idea is the same.

You'd Like This If You...

- Care deeply about nature
- Want to protect people and animals
- Like solving real-world problems
- Enjoy science with a purpose
- Want to make the world healthier

Why this matters to you:

Environmental engineering helps protect the air we breathe and the water we drink—things everyone needs to survive.

Environmental engineers don't just clean up messes.

They help build a better future for the planet.

Chemical Engineer
From Shampoo to Space Fuel

Your shampoo foams just right.
Medicine works safely inside your body.
Rocket fuel burns with powerful energy.
All of those things are made from chemicals–mixed carefully and on purpose.
If you like mixing, experimenting, and wondering what happens when things combine, you might already be thinking like a **chemical engineer.**

What They Do

Chemical engineers figure out how to turn raw materials into useful products.

They don't just mix chemicals for fun.

They design safe, controlled systems that make things like:

- Shampoo and soap
- Medicine
- Plastics and fabrics
- Food ingredients
- Fuels and batteries

Chemical engineers ask:

- How do we make this safely?
- How can we make it faster or cleaner?
- How do we avoid waste or danger?
-

Their work helps turn science ideas into everyday products.

Where You See Their Work in Real Life

You see chemical engineers' work when:

- Shampoo smells nice and rinses clean
- Medicine is pure and safe
- Plastics are shaped into useful items
- Fuel powers cars, planes, and rockets
- Food stays fresh longer

From bathrooms to space missions, chemical engineers are involved.

Cool Tools & Tech

Chemical engineers use tools like:

- Reactors and mixing systems
- Pipes and tanks that control flow
- Sensors that track temperature and pressure
- Computer simulations
- Safety equipment

Wow fact:

👉 A tiny change in temperature or pressure can completely change how a chemical reaction behaves.
That's why chemical engineers are very precise.

Future Problem They Help Solve

Chemical engineers help solve problems like:

- Making cleaner fuels
- Creating biodegradable plastics
- Producing medicine faster
- Reducing industrial pollution
- Improving batteries and energy storage

They help make products safer for people and the planet.

Try It Now: Why Does It Fly? (Observe)

You'll need:

- Baking soda
- Vinegar
- A clear cup

Steps:

1. Put a spoon of baking soda in the cup.
2. Slowly add vinegar.
3. Watch the bubbles form.

That bubbling is a chemical reaction—something chemical engineers study and control on a much larger scale.

You'd Like This If You...

- Enjoy experiments
- Like chemistry and science
- Are careful and detail-focused
- Want to create useful products
- Enjoy solving process problems

Why this matters to you:

Chemical engineering helps turn science into safe, useful things people use every day.

Chemical engineers don't just mix chemicals.

They design how the world makes things.

Industrial Engineer
Makes Systems Run Smoothly

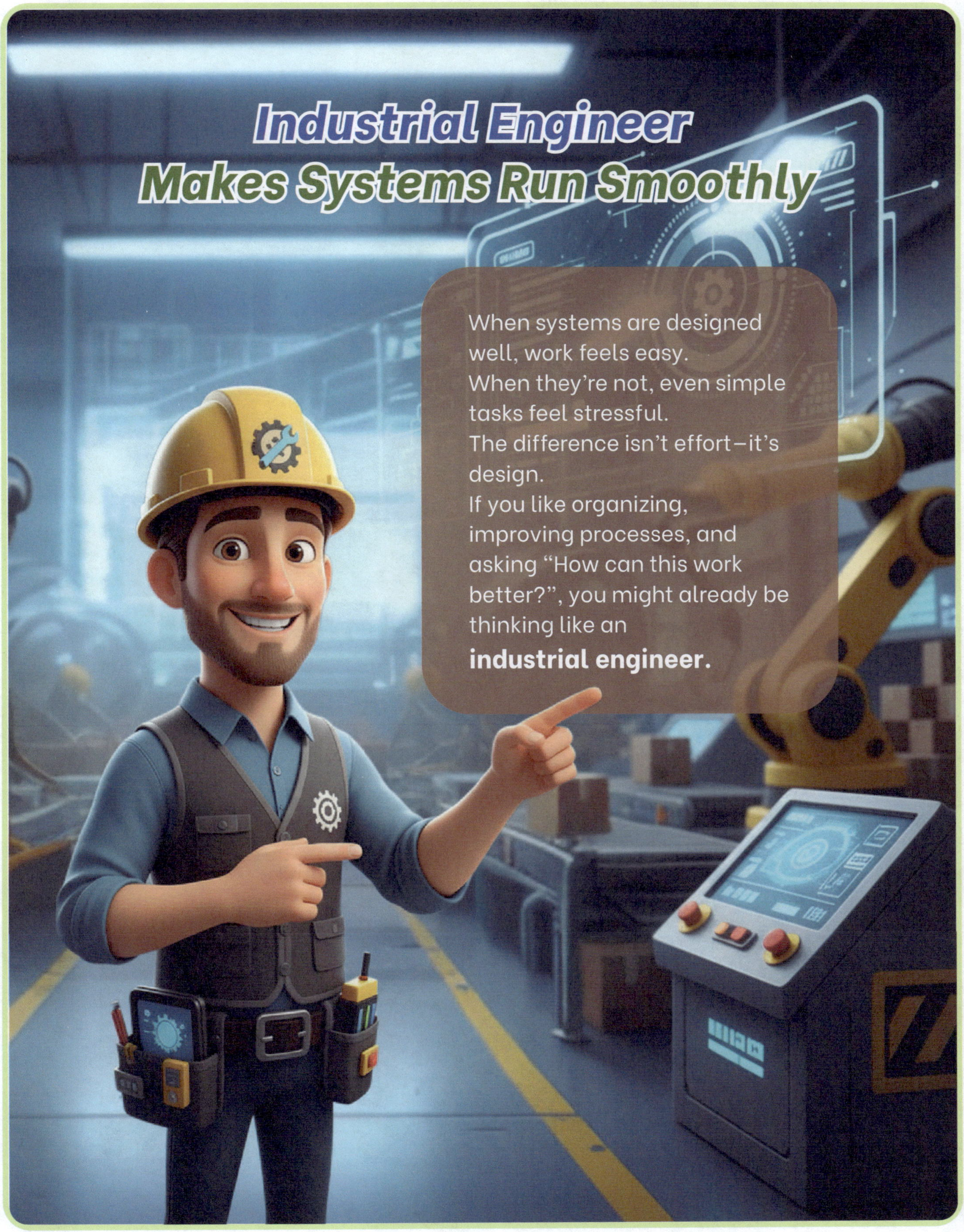

When systems are designed well, work feels easy.
When they're not, even simple tasks feel stressful.
The difference isn't effort–it's design.
If you like organizing, improving processes, and asking "How can this work better?", you might already be thinking like an **industrial engineer.**

What They Do

Industrial engineers focus on how systems work.
They don't design one machine or product.
They design how people, tools, and steps fit together.
They work on:

- Factories and production lines
- Hospitals and schools
- Transportation systems
- Warehouses and delivery routes
- Businesses and organizations

Their goal is to:

- Save time
- Reduce waste
- Improve safety
- Make work easier for people

They turn messy systems into smooth ones.

Where You See Their Work in Real Life

You see industrial engineers' work when:

- Lines move faster
- Workspaces feel organized
- Deliveries arrive on time
- Hospitals reduce waiting times
- Schools improve schedules

If something feels efficient instead of stressful, an industrial engineer likely helped.

Cool Tools & Tech

Industrial engineers use tools like:

- Flowcharts to map steps
- Spreadsheets and data tools
- Timing and measurement tools
- Simulation software
- Problem-solving methods

Wow fact:

👉 A small change—like moving a table or changing a step—can save thousands of hours over time.

Future Problem They Help Solve

Industrial engineers help solve problems like:

- Reducing waste in factories
- Improving safety at work
- Making healthcare more efficient
- Designing smarter cities
- Saving energy and resources

They help systems work better for both people and the planet.

Try It Now: Make It Smoother (Reflect)

You'll need:

- *Paper*
- *Pencil*

Steps:

1. Think of a daily routine (getting ready, packing a bag, lining up).
2. Write the steps in order.
3. Ask: Which step could be moved, removed, or improved to save time?

👉 Industrial engineers focus on making systems work better for people.

You'd Like This If You...

- Like organizing and planning
- Enjoy problem-solving
- Notice inefficiencies
- Care about fairness and ease
- Want to improve how things run

Why this matters to you:

Industrial engineering helps make everyday systems less stressful and more effective.

Industrial engineers don't just fix machines.

They fix how work happens.

Automotive Engineer
Future Cars & EVs

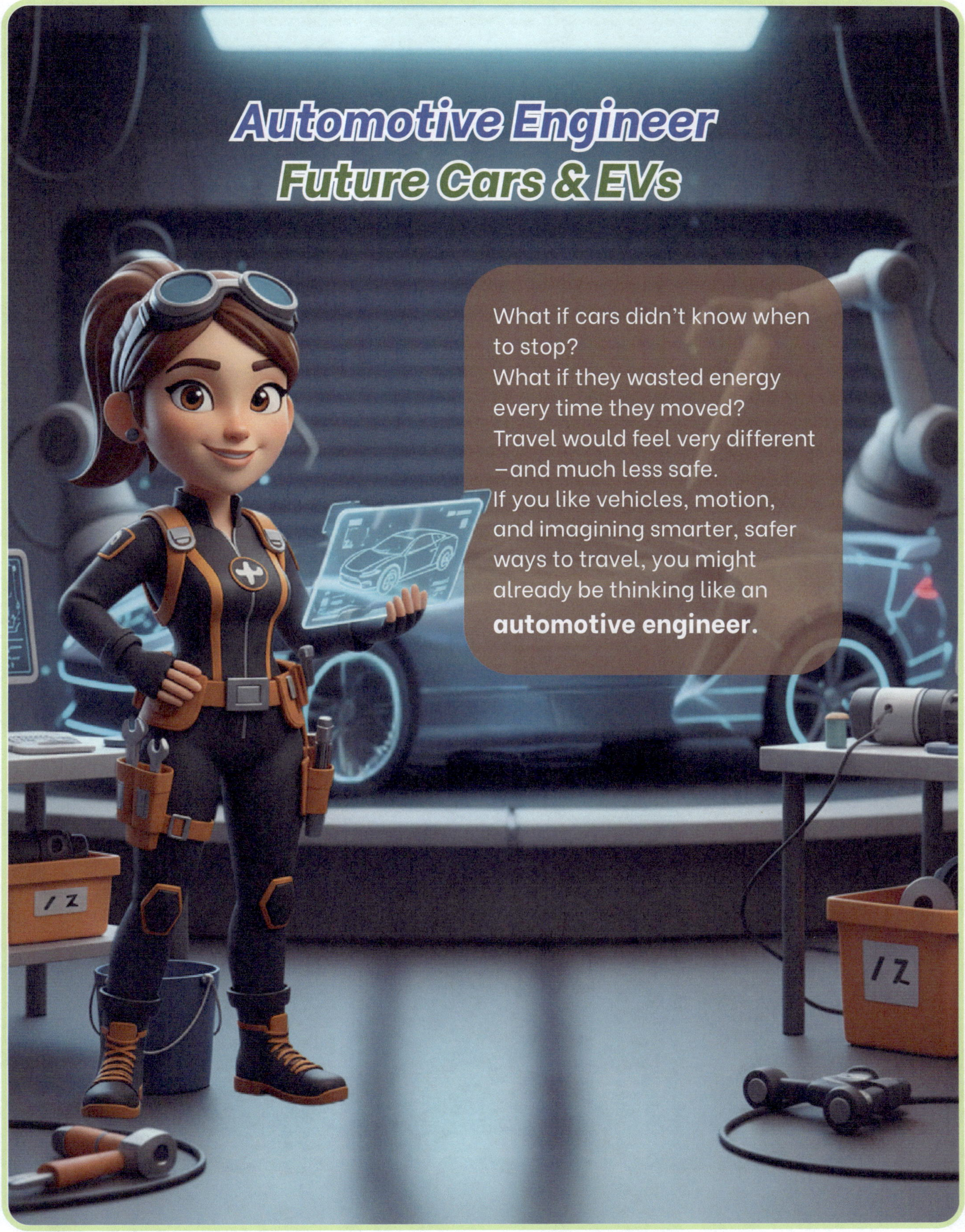

What if cars didn't know when to stop?
What if they wasted energy every time they moved?
Travel would feel very different —and much less safe.
If you like vehicles, motion, and imagining smarter, safer ways to travel, you might already be thinking like an **automotive engineer.**

What They Do

Automotive engineers design and improve vehicles.

They work on:

- Cars, trucks, and buses
- Electric vehicles (EVs)
- Engines and motors
- Safety systems
- Vehicle electronics

They think about:

- How vehicles move
- How they stop
- How to keep passengers safe
- How to use less energy

Modern automotive engineers help create cleaner, smarter transportation.

Where You See Their Work in Real Life

You see automotive engineers' work when:

- Electric cars charge instead of using gas
- Airbags protect passengers
- Braking systems prevent skidding
- Cars use sensors to assist drivers
- Vehicles become quieter and cleaner

Anytime a car improves, automotive engineers help make it happen.

Cool Tools & Tech

Automotive engineers use tools like:

- Computer simulations of crashes and airflow
- Electric motors and batteries
- Sensors and cameras
- Wind tunnels
- Test tracks

Wow fact:

👉 Electric vehicles have fewer moving parts than gas cars—which can mean less maintenance.

Future Problem They Help Solve

Automotive engineers help solve problems like:

- Reducing air pollution
- Making roads safer
- Improving battery life
- Designing self-driving features
- Creating transportation for the future

They help decide how people will move in the years to come.

Try It Now: Design a Future Car

You'll need:

- Paper
- Pencil

Steps:

1. Draw a car for the future.
2. Decide how it's powered (electric, solar, something new).
3. Add one safety feature and one eco-friendly feature.

That's how new vehicle ideas begin.

You'd Like This If You...

- Love cars or vehicles
- Enjoy engineering and design
- Care about safety and the environment
- Like solving motion problems
- Want to shape the future of travel

Why this matters to you:

Automotive engineering helps create safer, cleaner ways for people to move every day.

Automotive engineers don't just build cars.

They build the future of transportation.

Sound Engineer
Music, Movies, and More

Music is too loud and hurts your ears.
Voices are too quiet to understand.
One small mistake can ruin how sound feels.
Sound doesn't just happen.
Someone carefully shapes it.
If you like music, movies, podcasts, or noticing how sound feels in different places, you might already be thinking like a
sound engineer.

What They Do

Sound engineers work with sound waves.
They design, record, mix, and control sound so it:

- Sounds clear
- Feels balanced
- Matches the moment

They work on:

- Music recordings
- Concerts and live shows
- Movies and TV shows
- Video games
- Podcasts and audiobooks

Sound engineers decide:

- How loud something should be
- Where sound comes from
- How voices and music blend together
- How to remove noise and improve clarity

Their job is to make the sound feel right.

Where You See Their Work in Real Life

You see sound engineers' work when:

- Songs sound clean through headphones
- Dialogue is clear in a movie
- Video games feel immersive
- Concerts sound powerful but controlled
- Online videos don't hurt your ears

If the sound feels smooth instead of messy, a sound engineer helps.

Cool Tools & Tech

Sound engineers use tools like:

- Microphones to capture sound
- Mixing boards with sliders and knobs
- Audio software to edit sound
- Speakers and headphones
- Sound meters to measure volume

Wow fact:

👉 Sound travels as waves—and different shapes of rooms can completely change how sound feels.

Future Problem They Help Solve

Sound engineers help solve problems like:

- Making concerts safer for hearing
- Improving sound for virtual reality
- Helping people with hearing loss
- Designing better communication systems
- Creating more realistic games and movies

Sound engineering mixes science, creativity, and technology.

Try It Now: Sound Explorer

You'll need:

- A quiet room
- Different objects (cup, book, table)

Steps:

1. Tap each object gently.
2. Listen to how the sound changes.
3. Try tapping harder or softer.

Sound engineers study how materials change sound just like this.

You'd Like This If You...

- Love music or movies
- Notice sound details
- Enjoy technology and creativity
- Like experimenting
- Want to shape how people hear the world

Why this matters to you:

Sound shapes how we feel, learn, and connect—and sound engineers help make those experiences better.

Sound engineers don't just hear sound.

They design how the world sounds.

PART 4

Math Magicians

Finding Meaning Inside Numbers

Math isn't just about homework problems.

It's about finding patterns, making predictions, and figuring out what's really going on.

Math magicians use numbers to answer big questions:

- What's likely to happen next?
- What's the safest choice?
- What's fair?
- What's the best strategy?

They don't just calculate.
They think.

Some work with charts and graphs.
Some work with puzzles and strategies.
Some help doctors, scientists, and planners make smart decisions.

As you explore these careers, notice what clicks for you:

- Patterns
- Logic
- Probability
- Strategy
- Explaining ideas clearly

Math is a tool—and in the right hands, it becomes powerful.
Let's meet the people who use numbers to make sense of the world.

Statistician
Truth Finder in Numbers

Two people argue about what's "usually true."
One says, "Most kids like this."
The other says, "No, they don't."
You don't guess.
You ask, "What do the numbers say?"
If you like facts, patterns, and finding the real story behind lots of information, you might already be thinking like a **statistician.**

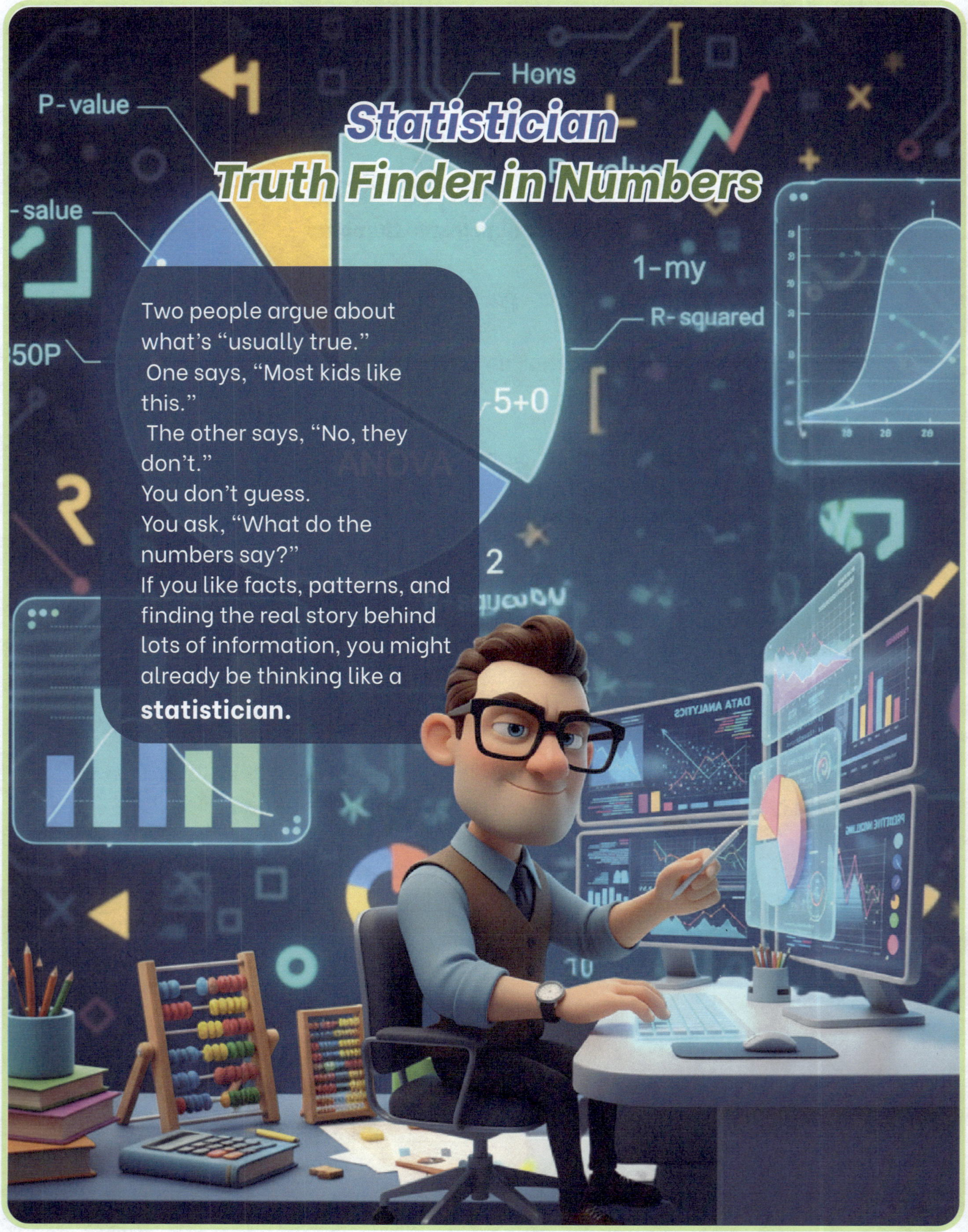

What They Do

Statisticians study data to find out what is actually happening.
Data is information collected from the world, like:

- Test scores
- Weather reports
- Sports stats
- Survey answers
- Medical results

Statisticians help answer questions like:

- What's common and what's rare?
- Is this result real—or just luck?
- What usually happens, and what doesn't?
- How confident can we be about this answer?

They don't just look at one number.
They look at patterns across many numbers.

Where You See Their Work in Real Life

You see statisticians' work when:

- Doctors test whether a medicine works
- Weather forecasts predict rain chances
- Sports teams analyze player performance
- Schools study learning progress
- News reports share polls and surveys

Whenever decisions are based on data instead of guesses, a statistician helps.

Cool Tools & Tech

Statisticians use tools like:

- Charts and graphs
- Spreadsheets and databases
- Statistical formulas
- Computer programs
- Probability models

Wow fact:

👉 A small sample of data can sometimes tell the truth about a very large group—if it's collected the right way.

Future Problem They Help Solve

Statisticians help solve problems like:

- Understanding disease trends
- Improving education systems
- Making fair decisions using data
- Predicting risks and outcomes
- Spotting misleading information

They help protect people from being fooled by bad data.

Try It Now: Mini Data Check (Observe)

You'll need:

- Paper
- Pencil

Steps:

1. Ask 5 people a simple yes/no question (for example: "Do you like pizza?").
2. Count how many say yes and how many say no.
3. Ask: Is this enough information to speak for everyone? Why or why not?

👉 Statisticians use data to move from opinions to evidence.

You'd Like This If You...

- Like facts and fairness
- Enjoy charts and patterns
- Ask "How do we know?"
- Care about truth over opinions
- Want numbers to mean something

Why this matters to you:

Statistics help the world make smarter, fairer decisions—and help people understand what's really going on.

Statisticians don't just count numbers.

They find the truth hiding inside them.

Actuary – Risk Calculator (Keeps Things Safe)

A company promises to help people after accidents.
A city plans for floods that might happen years from now.
A school wants to know how to stay safe in emergencies.
These choices aren't guesses. They're based on careful thinking about risk.
If you like thinking ahead, weighing possibilities, and using numbers to plan for the future, you might already be thinking like an **actuary.**

What They Do

Actuaries use math to predict risk.

Risk means:

- How likely something is to happen
- How big could the impact be

Actuaries study things like:

- Accidents
- Natural disasters
- Health problems
- Lifespans
- Money and costs

They help organizations decide:

- What plans are safe
- How much to prepare
- How to protect people fairly

Their goal is not to scare people—but to keep systems stable and safe.

Where You See Their Work in Real Life

You see actuaries' work when:

- Insurance plans exist
- Safety rules are created
- Companies prepare for emergencies
- Cities plan for disasters
- Long-term programs stay funded

If something keeps working even after something bad happens, an actuary likely helped plan for it.

Cool Tools & Tech

Actuaries use tools like:

- Probability models
- Charts and simulations
- Large data sets
- Computer programs
- Scenario testing

Wow fact:

👉 Actuaries often plan for events that might not happen for decades.

Future Problem They Help Solve

Actuaries help solve problems like:

- Planning for climate risks
- Preparing healthcare systems
- Making long-term plans fair
- Protecting people financially
- Helping societies handle uncertainty

They help systems survive the unexpected.

Try It Now: Risk Thinking Game

You'll need:

- Paper
- Pencil

Steps:

1. List three things that could go wrong during a school event.
2. Rank them from least likely to most likely.
3. Write one way to prepare for each.

That's how actuaries think—planning calmly instead of panicking.

You'd Like This If You...

- Like thinking ahead
- Enjoy probability and logic
- Care about safety and fairness
- Prefer calm planning
- Want to protect people using math

Why this matters to you:

Actuaries help keep systems steady—even when the future is uncertain.

Actuaries don't predict the future.

They help us be ready for it.

Operations Research Analyst
Big Problem Puzzle Solver

A city needs to get buses to schools faster.

A hospital wants to help more patients with the same number of doctors.

A delivery company needs to choose the fastest routes—every single day.

There are many choices, and choosing the wrong one wastes time, money, or energy.

If you love puzzles with more than one right answer and enjoy finding the best solution, you might already be thinking like an **operations research analyst**.

What They Do

Operations research analysts use math, logic, and data to solve very big problems.
These problems usually involve:

- Many steps
- Many people
- Limited time or resources

They ask questions like:

- What's the smartest way to do this?
- What should happen first?
- Where do delays come from?
- How can we do more with less?

Instead of guessing, they build models—simplified versions of real-life systems—to test different choices safely.
Think of it like playing a strategy game... but for real life.

Where You See Their Work in Real Life

You see operations research analysts' work when:

- Delivery trucks arrive on time
- Airlines schedule flights efficiently
- Hospitals reduce waiting times
- Factories avoid bottlenecks
- Cities improve traffic flow

Anytime a system runs smoother than expected, someone solved a big puzzle behind the scenes.

Cool Tools & Tech

Operations research analysts use tools like:

- Flowcharts and diagrams
- Spreadsheets and simulations
- Optimization models
- Computer program
- Decision trees

Wow fact:

👉 Sometimes the "best" solution isn't the fastest—it's the one that saves the most energy or helps the most people.

Future Problem They Help Solve

Operations research analysts help solve problems like:

- Reducing traffic and pollution
- Making emergency response faster
- Improving school schedules
- Using resources more fairly
- Planning smart cities

They help systems work better—not harder.

Try It Now: The Best Route Challenge

You'll need:

- Paper
- Pencil

Steps:

1. Draw a map with 5 locations (home, school, store, park, library).
2. Draw different routes between them.
3. Try to find the route that visits all places in the least distance.
4. Did you try more than one solution?

That's exactly what operations research is about.

You'd Like This If You...

- Love strategy games and puzzles
- Enjoy thinking through many options
- Like logic and planning
- Care about efficiency and fairness
- Want to solve real-world problems

Why this matters to you:

Big systems affect millions of people—and smart planning can make life easier for everyone.

Operations research analysts don't rush to answers.

They find the smartest path forward.

Cryptographer
Code Maker & Code Breaker

You send a secret message.
Only one person can read it.
Even if someone else sees it,
it makes no sense to them.
That's not magic.
That's math protecting
information.
If you love puzzles, secret
messages, and thinking
about how to keep
information safe, you might
already be thinking like a
cryptographer.

What They Do

Cryptographers create and study codes.

Their job is to:

- Lock information so only the right people can read it
- Unlock coded messages when allowed
- Make sure messages aren't changed along the way

They work with:

- Numbers
- Patterns
- Logic
- Algorithms (step-by-step rules)

Cryptography helps protect:

- Messages
- Passwords
- Photos
- Money
- Important information

A cryptographer's goal is simple but serious:
Keep secrets safe.

Where You See Their Work in Real Life

You see cryptographers' work when:

- A website shows a lock icon
- You send a private message
- You log in with a password
- Online payments stay secure
- Apps protect your personal data

Every time your information stays private online, cryptography is working quietly in the background.

Cool Tools & Tech

Cryptographers use tools like:

- Number theory (a special kind of math)
- Algorithms to lock and unlock data
- Large prime numbers
- Computers to test code strength
- Encryption systems

Wow fact:

👉 Some modern codes would take thousands of years for computers to break—if they're designed well.

Future Problem They Help Solve

Cryptographers help solve problems like:

- Keeping online communication private
- Protecting money and digital wallets
- Securing medical and school records
- Preparing for future computers that are much faster
- Preventing digital spying and fraud

As technology grows, cryptography must grow with it.

Try It Now: Make Your Own Secret Code

You'll need:

- Paper
- Pencil

Steps:

1. Write the alphabet in a line.
2. Write it again, but start from a different letter (like starting A as D).
3. Replace each letter in your message with the matching letter below.
4. Can a friend crack your code?

That's how cryptography begins—simple ideas that grow stronger over time.

You'd Like This If You...

- Love puzzles and logic
- Enjoy secret messages
- Like math with a purpose
- Care about privacy and fairness
- Want to protect information

Why this matters to you:

Cryptography helps keep people safe in a digital world—where information is powerful.

Cryptographers don't shout about their work.

They quietly guard secrets that matter.

Financial Quant
Math for Money & Markets

Prices go up.
Prices go down.
Markets change every second.
Most people see confusion.
You see patterns.
If you like numbers, trends, and asking “What’s likely to happen next?”, you might already be thinking like a **financial quant.**

What They Do

Financial quants (short for quantitative analysts) use math and data to understand money systems.

They study things like:

- Prices
- Risks
- Patterns over time
- How choices affect outcomes

They build models that help answer questions like:

- Is this a risky decision?
- What's the safest option?
- How do many small changes add up?

Quants don't guess.
They use numbers to make careful predictions.

Where You See Their Work in Real Life

Financial quants (short for quantitative analysts) use math and data to understand money systems.

They study things like:

- Prices
- Risks
- Patterns over time
- How choices affect outcomes

They build models that help answer questions like:

- Is this a risky decision?
- What's the safest option?
- How do many small changes add up?

Quants don't guess.
They use numbers to make careful predictions.

Cool Tools & Tech

Financial quants use tools like:

- Advanced math formulas
- Large data sets
- Computer simulations
- Graphs and models
- Programming tools

Wow fact:

👉 Some financial models test millions of "what if?" situations before choosing a safe path.

Future Problem They Help Solve

Financial quants help solve problems like:

- Reducing financial crashes
- Managing risk during uncertain times
- Helping systems stay fair and stable
- Planning long-term investments
- Understanding complex economic behavior

They help decision-makers see risks before problems happen.

Try It Now: Predict the Trend

You'll need:

- Paper
- Pencil

Steps:

1. Write down the prices of an item you see often over a week.
2. Make a simple chart.
3. Guess what might happen next—and why.

That's how quants start spotting trends.

You'd Like This If You...

- Enjoy math and logic
- Like patterns over time
- Think carefully before deciding
- Want numbers to guide choices
- Enjoy solving complex puzzles

Why this matters to you:

Financial math helps systems stay balanced—and helps people make smarter choices with resources.

Financial quants don't chase money.

They study how systems behave—and help keep them stable.

Epidemiologist
Disease Pattern Tracker

A few people get sick in one place. Then a few more in another place. Someone notices it's not random. That's where the tracking begins. If you like patterns, care about people's health, and want to stop problems before they grow bigger, you might already be thinking like an **epidemiologist.**

What They Do

Epidemiologists study how diseases spread.
They don't usually treat patients directly.
Instead, they ask big-picture questions like:

- Where did this illness start?
- How is it spreading?
- Who is most at risk?
- What can stop it?

They collect and study data about:

- Who gets sick
- When symptoms appear
- How people interact
- How fast do illnesses move

Their goal is to protect communities, not just individuals.

Where You See Their Work in Real Life

You see epidemiologists' work when:

- Health officials explain outbreaks
- Schools close or reopen for safety
- Vaccination plans are created
- Doctors get guidelines during flu season
- News talks about "flattening the curve."

Anytime a health decision is made for many people at once, epidemiologists help guide it.

Cool Tools & Tech

Epidemiologists use tools like:

- Charts and graphs showing cases over time
- Maps that track where illnesses spread
- Databases of health information
- Math models that simulate outbreaks
- Computer software to test "what if" scenarios

Wow fact:

👉 Some disease models can show how one small change–like washing hands more–can protect thousands of people.

Future Problem They Help Solve

Epidemiologists help solve problems like:

- Stopping outbreaks early
- Protecting vulnerable populations
- Planning responses to new diseases
- Improving public health systems
- Helping countries work together during emergencies

They help turn information into prevention.

Try It Now: Spread Simulation

You'll need:

- Paper
- Pencil

Steps:

1. Draw 10 stick figures in a group.
2. Color one figure red to represent illness.
3. Each "day," draw lines to show who interacts.
4. Color new figures red when they connect.

What happens when people interact less?
That's how epidemiologists study spread.

You'd Like This If You...

- Care about helping people
- Like patterns and logic
- Enjoy big-picture thinking
- Want to prevent problems early
- Like science with real impact

Why this matters to you:

Epidemiology helps protect entire communities by understanding how diseases move—and how to stop them.

Epidemiologists don't wait for problems to explode.

They track, predict, and protect.

Logistics Planner
Delivery & Supply Wizard

A package arrives exactly on time.
A grocery store never runs out of food.
Hospitals get supplies when they need them most.
Behind all of that is careful planning.
If you like organizing, mapping routes, and figuring out the smartest way to move things from one place to another, you might already be thinking like a **logistics planner.**

What They Do

Logistics planners figure out how things move.

They plan how to:

- Deliver food, clothes, and supplies
- Move items from factories to stores
- Get medicine to hospitals quickly
- Keep shelves stocked without waste

They ask questions like:

- What's the fastest route?
- What's the safest way?
- How do we avoid delays?
- What happens if something goes wrong?

Their job is to make sure the right things reach the right place at the right time.

Where You See Their Work in Real Life

You see logistics planners' work when:

- Online orders arrive on schedule
- Stores stay stocked
- School supplies show up before classes start
- Emergency supplies reach people quickly
- Food stays fresh during transport

If things arrive smoothly instead of being late or missing, a logistics planner helps.

Cool Tools & Tech

Logistics planners use tools like:

- Maps and route planners
- Tracking systems
- Spreadsheets and schedules
- Simulation software
- Communication systems

Wow fact:

👉 A tiny delay in one place can affect deliveries all over the world—which is why logistics planners plan backup routes.

Future Problem They Help Solve

Logistics planners help solve problems like:

- Reducing wasted food and materials
- Making deliveries more eco-friendly
- Helping during disasters and emergencies
- Supporting global trade
- Keeping supplies moving during crises

They help keep the world connected.

Try It Now: Delivery Route Challenge

You'll need:

- Paper
- Pencil

Steps:

1. Draw a map with a warehouse and 5 delivery stops.
2. Plan the fastest route to visit all stops.
3. Try a second route and compare.

Which one worked better?
That's exactly how logistics planners test ideas.

You'd Like This If You...

- Enjoy organizing and planning
- Like maps and routes
- Care about efficiency
- Think ahead about problems
- Want systems to work smoothly

Why this matters to you:

Logistics planning helps make sure people get what they need—especially when it matters most.

Logistics planners don't just move things.

They keep life moving.

Game Theorist

Strategy Scientist

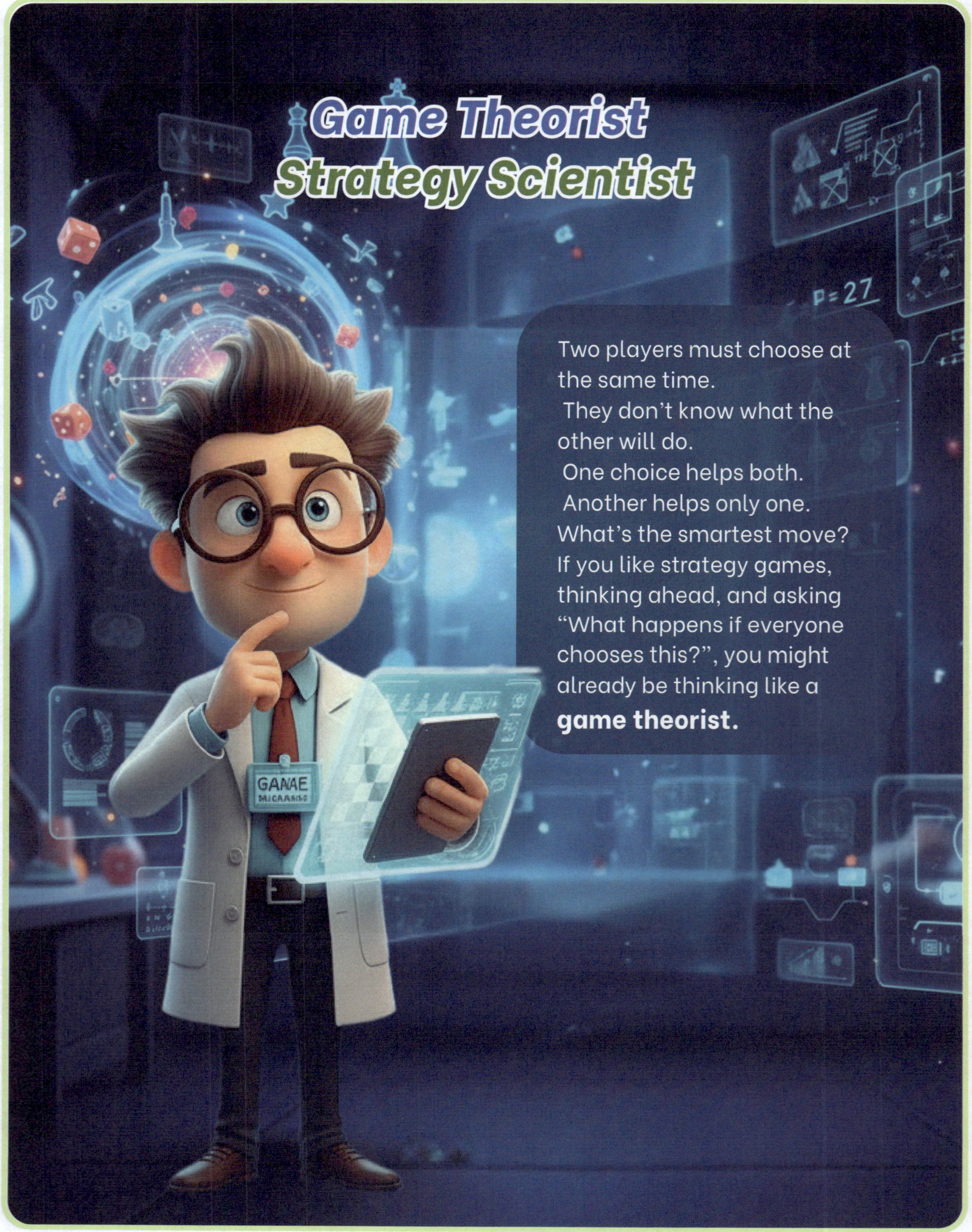

Two players must choose at the same time.
They don't know what the other will do.
One choice helps both.
Another helps only one.
What's the smartest move?
If you like strategy games, thinking ahead, and asking "What happens if everyone chooses this?", you might already be thinking like a **game theorist.**

What They Do (Kid Version)

Game theorists study decision-making.
Not video games but real-life situations where:

- People make choices
- Choices affect others
- Outcomes depend on everyone's decisions

They study questions like:

- When should you cooperate?
- When should you compete?
- What's fair for everyone?
- What happens if someone cheats?

Game theorists use math and logic to understand how decisions interact.

Where You See Their Work in Real Life

You see game theory at work when:

- Teams decide whether to work together
- Countries negotiate rules or treaties
- Businesses set prices
- Online games balance fairness
- Schools design group activities

Anytime choices affect more than one person, game theory helps explain what might happen next.

Cool Tools & Tech

Game theorists use tools like:

- Decision trees
- Payoff tables (what each choice leads to)
- Probability and logic
- Computer simulations
- Strategic models

Wow fact:

👉 Some strategies that seem "smart" in the short term can cause everyone to lose in the long term.

Future Problem They Help Solve

Game theorists help solve problems like:

- Encouraging cooperation instead of conflict
- Designing fair rules
- Preventing cheating in systems
- Understanding competition
- Helping groups make better decisions

They help people see beyond "me" and think about "us."

Try It Now: The Choice Game

You'll need:

- A friend
- Paper

Steps:

1. Each of you secretly chooses A or B.
2. Reveal choices at the same time.
3. Decide rewards for each outcome (both choose A, one chooses B, etc.).

What strategy worked best over many rounds?
That's how game theorists test ideas.

You'd Like This If You...

- Love strategy games
- Enjoy thinking ahead
- Care about fairness
- Like logic and problem-solving
- Want to understand how people make choices

Why this matters to you:

Understanding strategy helps people work together, compete fairly, and avoid bad outcomes.

Game theorists don't tell people what to do.

They help explain what happens when choices collide.

Math Teacher / Communicator Makes Math Make Sense

A math problem finally clicks.
Numbers stop feeling scary.
You hear someone say,
"Ohhh, now I get it."
That moment matters.
If you like helping others understand things, explaining ideas in simple ways, and turning confusion into confidence, you might already be thinking like a **math teacher or math communicator.**

What They Do

Math teachers and communicators help people understand math, not just memorize it.

They don't ask:

- "Can you get the answer fast?"

They ask:

- "Do you understand why this works?"

They help learners:

- Break big problems into smaller steps
- See patterns instead of just numbers
- Use math in real life
- Build confidence instead of fear

Some work in classrooms.
Some create books, videos, games, or online lessons.
Some teach kids. Some teach adults.
Their real job is to help people believe,
"I can learn this."

Where You See Their Work in Real Life

You see math communicators' work when:

- A teacher explains a hard idea in a new way
- A video makes math feel simple
- A book shows math through stories or pictures
- A game teaches math without feeling like homework
- Someone helps you understand mistakes instead of judging them

Anytime math feels clearer and kinder, a math communicator helped.

Cool Tools & Tech

Math teachers and communicators use tools like:

- Visual models (drawings, blocks, charts)
- Games and puzzles
- Stories and real-life examples
- Videos and interactive tools
- Questions that guide thinking

Wow fact:

👉 The human brain understands math better when it's connected to pictures, stories, or movement.

Future Problem They Help Solve

Math teachers and communicators help solve problems like:

- Math anxiety and fear
- Learning gaps in schools
- Making STEM more accessible
- Helping kids think logically
- Teaching problem-solving, not just answers

They help build thinkers—not just test-takers.

Try It Now: Teach It Back

You'll need:

- A friend or family member

Steps:

1. Pick a math idea you understand (adding fractions, patterns, percentages).
2. Explain it using only simple words or drawings.
3. Ask them what part made sense—and what didn't.

Teaching others helps you learn better, too.

You'd Like This If You...

- Enjoy helping others
- Like explaining ideas
- Are patient and encouraging
- Believe learning should feel safe
- Want to change how people feel about math

Why this matters to you:

Understanding math opens doors—and helping others understand it opens even more.

Math teachers and communicators don't just teach numbers.

They help people believe in their own thinking.

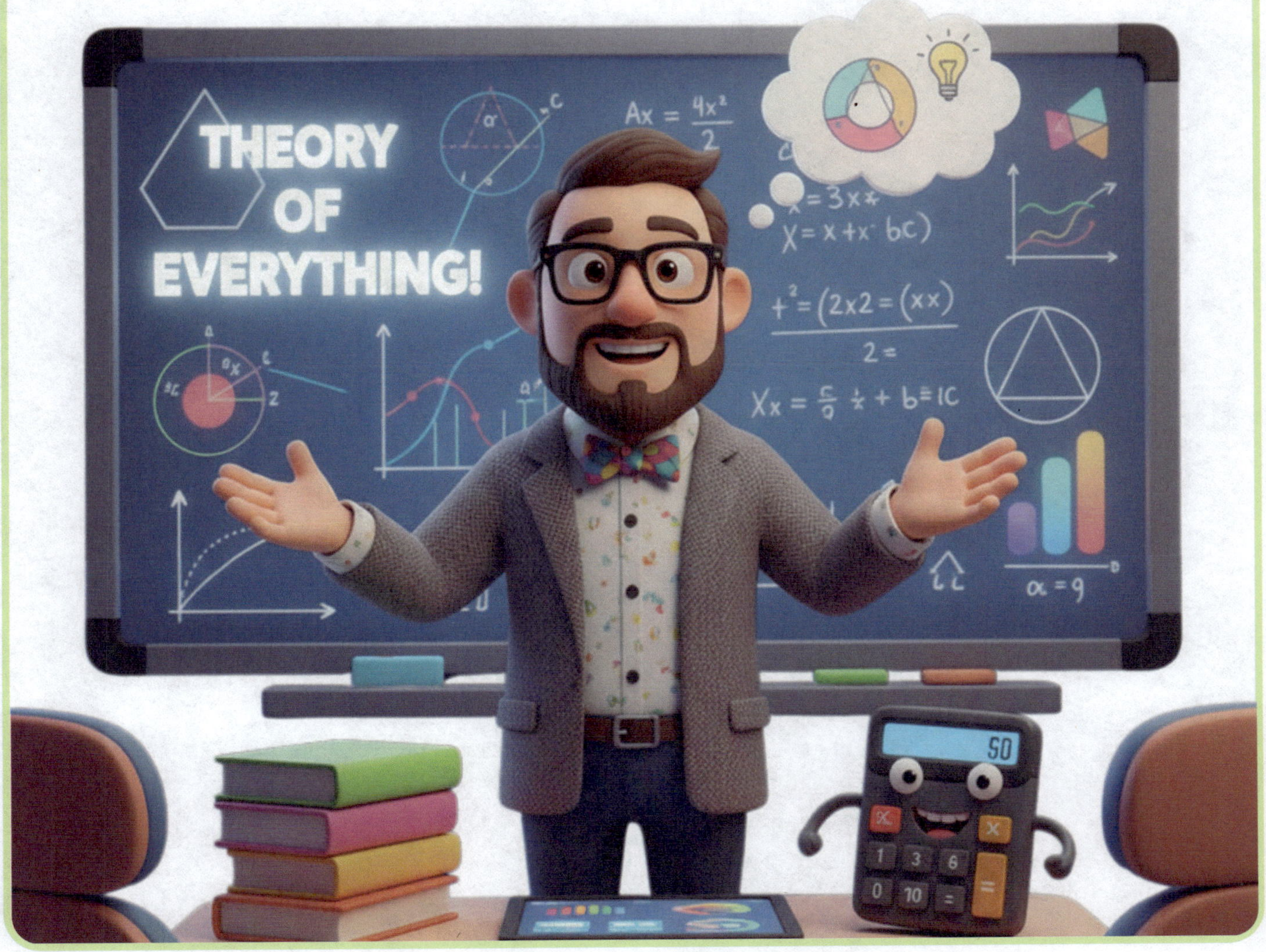

Astronomer
Universe Mathematician

You look up at the night sky.
Thousands of tiny lights stare back.
Some are stars.
Some are planets.
Some are galaxies so far away that their light began traveling before humans existed.
If you feel curious instead of small when you look at space, you might already be thinking like an **astronomer.**

What They Do (Kid Version)

Astronomers study everything beyond Earth.

They ask questions like:

- What are stars made of?
- How do planets move?
- Are there other worlds like ours?
- How did the universe begin?

Astronomers don't travel to space most of the time.
They use math, physics, and light to understand what's happening far away.
They turn tiny dots of light into real information.

Where You See Their Work in Real Life

You see astronomers' work when:

- Telescopes capture images of planets and galaxies
- Space missions send back data
- Scientists explain eclipses or meteor showers
- Calendars predict moon phases
- News talks about new planets discovered

Even your phone's GPS depends on space science.

Cool Tools & Tech

Astronomers use tools like:

- Telescopes on Earth and in space
- Cameras that detect invisible light
- Computers that analyze huge data sets
- Math models to predict motion
- Satellites and space probes

Wow fact:

👉 Some stars you see tonight may have already burned out—but their light is still traveling toward us.

Future Problem They Help Solve

Astronomers help solve problems like:

- Understanding Earth's place in the universe
- Tracking dangerous asteroids
- Studying space weather that affects satellites
- Learning how planets form
- Exploring whether life exists elsewhere

Their discoveries often lead to new technology back on Earth.

Try It Now: Sky Watcher

You'll need:

- A notebook
- A clear night sky

Steps:

1. Pick one bright star or planet.
2. Sketch its position relative to the moon or trees.
3. Check again a few nights later.

Did it move?
That's how astronomers track objects—patiently, carefully.

You'd Like This If You...

- Love space and big questions
- Enjoy math and science together
- Are curious about the unknown
- Like observing patterns over time
- Feel excited by discovery

Why this matters to you:

Astronomy helps humans understand where we came from—and where we might go next.

Astronomers don't just study stars.

They study our place in the universe.

PART 5

Planet Protectors & Space Explorers

Caring for Earth – and Reaching Beyond It

Some STEM thinkers look down at Earth and ask,
"How do we protect this planet?"

Others look up at the sky and ask,
"What's out there?"

Planet protectors and space explorers do both.

They care deeply about:

- Clean air and water
- Healthy animals and ecosystems
- Safe cities and food systems
- Space, planets, and the future of humans beyond Earth

These careers mix science, engineering, and courage.
They solve problems that affect everyone, now and in the future.

As you read, notice what pulls you in:

- Protecting nature
- Designing clean energy
- Exploring oceans or space
- Helping communities prepare for disasters
- Inventing new ways to live better

This is where curiosity meets responsibility.
Let's meet the people helping protect our planet–and explore what comes next.

Renewable Energy Engineer
Solar & Wind Hero

Our house is powered by sunlight.
A windy hill creates electricity.
Cities run without filling the air with smoke.
Energy doesn't have to hurt the planet.
If you like building things and protecting nature, you might already be thinking like a **renewable energy engineer.**

What They Do (Kid Version)

Renewable energy engineers design systems that create clean power. They work with energy sources that don't run out, like:

- Sunlight (solar)
- Wind
- Water
- Heat from the Earth

They figure out:

- How to capture energy
- How to store it
- How to send it where it's needed

Their goal is to give people power without polluting the planet.

Where You See Their Work in Real Life

You see renewable energy engineers' work when:

- Solar panels sit on rooftops
- Wind turbines spin on hills or in the ocean
- Schools and homes use clean electricity
- Batteries store energy for later
- Cities reduce pollution

If you see energy coming from nature, an engineer helped design it.

Cool Tools & Tech

Renewable energy engineers use tools like:

- Solar panels and wind turbines
- Batteries and energy storage systems
- Sensors to track energy output
- Computer models to predict weather and power
- Electrical systems

Wow fact:

👉 One hour of sunlight reaching Earth could power the world for a whole year—if we could capture it all.

Future Problem They Help Solve

Renewable energy engineers help solve problems like:

- Reducing air pollution
- Fighting climate change
- Providing power to remote areas
- Making energy cheaper and cleaner
- Protecting natural resources
-

They help power the future responsibly.

Try It Now: Sun Tracker

You'll need:

- Paper
- Pencil

Steps:

1. Trace a shadow from an object in the morning.
2. Check again at midday and in the afternoon.
3. Notice how the sun moves.

That movement helps engineers decide where to place solar panels.

You'd Like This If You...

- Care about the environment
- Enjoy engineering and design
- Like clean energy ideas
- Want to help the planet
- Believe technology can do good

Why this matters to you:

Renewable energy helps protect Earth while powering modern life.

Renewable energy engineers don't just build systems.

They help protect the future.

Conservation Biologist
Animal & Habitat Protector

A forest stays alive for future generations.
An animal species avoids extinction.
A river becomes clean enough for life again.
Nature needs help–and some people choose to protect it.
If you love animals, care deeply about nature, and want to protect living things, you might already be thinking like a
conservation biologist.

What They Do (Kid Version)

Conservation biologists study plants, animals, and ecosystems to protect them.

They work to:

- Protect endangered species
- Restore damaged habitats
- Study how animals live and move
- Understand how human actions affect nature
-

They don't just study nature.

They help make plans to keep it alive.

Where You See Their Work in Real Life

You see conservation biologists' work when:

- Wildlife reserves are protected
- Animals are reintroduced into the wild
- Forests and reefs are restored
- Laws protect endangered species
- Nature documentaries teach conservation

If you see nature being protected, a conservation biologist was likely involved.

Cool Tools & Tech

Conservation biologists use tools like:

- GPS trackers on animals
- Cameras that watch wildlife
- Drones to study land
- Maps and data systems
- Field notebooks and samples

Wow fact:

👉 Some animals travel thousands of kilometers every year—and scientists track them to protect their paths.

Future Problem They Help Solve

Conservation biologists help solve problems like:

- Preventing animal extinction
- Protecting habitats from destruction
- Balancing human needs and nature
- Restoring ecosystems
- Helping wildlife adapt to climate change

They help protect Earth's living systems.

Try It Now: Habitat Observer

You'll need:

- Notebook
- Pencil

Steps:

1. Pick a small outdoor area (yard, park, street).
2. List all living things you see.
3. Think about what those living things need to survive.

That's how conservation biologists begin understanding habitats.

You'd Like This If You...

- Love animals and plants
- Enjoy being outdoors
- Care about protecting life
- Like science with purpose
- Want to defend nature

Why this matters to you:

Protecting nature helps protect all life—including humans.

Conservation biologists don't just study animals.

They stand up for them.

Urban Planner Designs Smart Cities

A city where parks are easy to reach.
Schools, homes, and shops are close together.
Streets feel safe for walking and biking.
Traffic flows instead of getting stuck.
Cities don't grow well by accident.
And if you like thinking about how people live together and how spaces affect daily life, you might already be thinking like an

urban planner.

What They Do

Urban planners design how cities work.
They think about:

- Where people live
- How they travel
- Where schools, parks, and hospitals go
- How to keep cities safe and healthy

Urban planners ask questions like:

- How can this city grow without chaos?
- Where should green spaces be?
- How do we reduce traffic and pollution?
- How can everyone access what they need?

They don't design one building.
They design how everything fits together.

Where You See Their Work in Real Life

You see urban planners' work when:

- Bike lanes connect neighborhoods
- Parks are built near homes
- Cities improve public transport
- New neighborhoods are designed
- Flood-prone areas are protected

If a city feels livable instead of stressful, an urban planner helps plan it.

Cool Tools & Tech

Urban planners use tools like:

- Maps and city models
- Data about population and traffic
- Computer simulations
- Community surveys
- Environmental impact tools

Wow fact:

👉 Some cities are designed so most daily needs are within a 15-minute walk.

Future Problem They Help Solve

Urban planners help solve problems like:

- Traffic congestion
- Pollution and noise
- Overcrowding
- Climate risks like flooding and heat
- Making cities fair and accessible

They help cities grow in smarter ways.

Try It Now: Design a Mini City

You'll need:

- Paper
- Pencil

Steps:

1. Draw a small city map.
2. Place homes, schools, parks, and roads.
3. Ask: Is everything easy to reach?

That's how urban planners start thinking.

You'd Like This If You...

- Care about people and communities
- Enjoy planning and design
- Like thinking big-picture
- Want cities to be fair and green
- Enjoy solving human problems

Why this matters to you:

Cities shape how people live, learn, and connect.

Urban planners don't just design cities.

They design daily life.

Agricultural Scientist
Future Food Grower

Farms grow more food using less water.
Crops survive heat, drought, and pests.
People around the world have enough to eat.
Feeding the planet takes science.
If you like plants, problem-solving, and helping people in a very real way, you might already be thinking like an **agricultural scientist.**

What They Do

Agricultural scientists study how we grow food.

They work on things like:

- Improving crops
- Protecting plants from disease
- Using water wisely
- Helping farmers grow more with less land
- Making food safer and healthier

They mix biology, chemistry, and environmental science to help feed a growing population.

Where You See Their Work in Real Life

You see agricultural scientists' work when:

- Crops grow stronger and healthier
- Farmers use smarter planting methods
- Food lasts longer without waste
- New fruits and vegetables appear
- Farming adapts to climate change

Every meal you eat is connected to agricultural science.

Cool Tools & Tech

Agricultural scientists use tools like:

- Soil testing kits
- Greenhouses and growth chambers
- Drones to monitor crops
- Data tools to track yields
- Lab equipment for plant research

Wow fact:

👉 Scientists are developing crops that need less water and fertilizer, helping protect the environment.

Future Problem They Help Solve

Agricultural scientists help solve problems like:

- Feeding more people with limited land
- Protecting crops from climate stress
- Reducing food waste
- Keeping soil healthy
- Making farming more sustainable

They help make sure food keeps growing—now and in the future.

Try It Now: Plant Scientist

You'll need:

- Seeds
- Soil
- Two small containers

Steps:

1. Plant seeds in both containers.
2. Give one more water or sunlight.
3. Observe which grows better—and why.

That's how agricultural science experiments begin.

You'd Like This If You...

- Like plants and nature
- Care about food and health
- Enjoy experiments
- Want to help people
- Like science with real impact

Why this matters to you:

Agricultural science helps feed the world while protecting the planet.

Agricultural scientists don't just grow food.

They grow in the future.

Disaster Scientist / Engineer
Earthquake & Storm Safety

A building stays standing during an earthquake.
A town gets a warning before a big storm.
People know where to go—and what to do—when danger comes.
Disasters can't always be stopped.
But the damage can be reduced.
If you care about protecting people and like planning for emergencies, you might already be thinking like a **disaster scientist or disaster engineer.**

What They Do

Disaster scientists and engineers study natural disasters to help keep people safe.

They work on problems like:

- Earthquakes
- Hurricanes and typhoons
- Floods
- Wildfires
- Landslides

They ask:

- Why do disasters happen?
- What areas are most at risk?
- How can buildings be made safer?
- How can people prepare ahead of time?

Their goal is to turn knowledge into protection.

Where You See Their Work in Real Life

You see their work when:

- Buildings are designed to shake but not fall
- Early warning systems send alerts
- Evacuation routes are planned
- Emergency drills are practiced
- Maps show danger zones

If lives are saved during disasters, disaster scientists help.

Cool Tools & Tech

Disaster scientists and engineers use tools like:

- Sensors that detect earthquakes
- Weather satellites
- Computer simulations of disasters
- Risk maps
- Engineering models

Wow fact:

👉 Some earthquake warning systems can alert people seconds before shaking starts—enough time to get to safety.

Future Problem They Help Solve

Disaster scientists and engineers help solve problems like:

- Protecting cities from stronger storms
- Designing safer buildings
- Improving emergency responses
- Helping communities recover faster
- Reducing loss of life

They help people stay ready instead of scared.

Try It Now: Safety Planner

You'll need:

- Paper
- Pencil

Steps:

1. Think of one natural disaster common in your area.
2. Write three things people could do to stay safe.
3. Draw a simple safety plan.

That's how disaster planning begins.

You'd Like This If You...

- Want to protect people
- Like science with real-world impact
- Think ahead and stay calm
- Care about safety
- Want to help communities

Why this matters to you:

Preparing for disasters saves lives—and science makes preparation possible.

Disaster scientists don't stop disasters.

They help people survive them.

Space Engineer (Mission Specialist)
Rockets & Rovers

A rocket launches perfectly
no mistakes allowed.
A rover lands on another
planet and starts exploring.
A satellite orbits Earth,
sending back data every
second.
Space is beautiful—but it's
also very unforgiving.
If you like extreme
challenges, careful planning,
and technology that must
work the first time, you might
already be thinking like a
space engineer.

What They Do

Space engineers design, test, and run space missions.
Mission specialists focus on making sure:

- Rockets launch safely
- Spacecraft survive heat, cold, and pressure
- Rovers move correctly on other planets
- Astronauts and equipment stay safe

They work on:

- Rockets and spacecraft
- Satellites
- Mars rovers
- Space stations
- Mission plans and controls

Every detail matters. There are no quick fixes in space.

Where You See Their Work in Real Life

You see space engineers' work when:

- Rockets launch into orbit
- Rovers send pictures from Mars
- Satellites help with GPS and weather
- Space stations orbit Earth
- Scientists receive space data

Even everyday things—like maps and weather forecasts—depend on space engineers.

Cool Tools & Tech

Space engineers use tools like:

- Rocket engines
- Heat shields
- Navigation systems
- Simulation software
- Mission control systems

Wow fact:

👉 A spacecraft must survive temperatures from very hot sunlight to freezing darkness—sometimes in the same orbit.

Future Problem They Help Solve

Space engineers help solve problems like:

- Sending humans safely to the Moon or Mars
- Protecting satellites from space debris
- Exploring planets and asteroids
- Improving communication on Earth
- Discovering new knowledge about space

They help humans explore farther than ever before.

Try It Now: Mission Planner

You'll need:

- Paper
- Pencil

Steps:

1. Choose a destination (Moon, Mars, asteroid).
2. Decide the mission goal.
3. List three challenges the spacecraft must survive.

That's how real space missions are planned.

You'd Like This If You...

- Love space and exploration
- Enjoy engineering challenges
- Think carefully and precisely
- Like teamwork and planning
- Dream big but plan smart

Why this matters to you:

Space engineering helps humans explore, communicate, and learn beyond Earth.

Space engineers don't just look at space.

They make reaching it possible.

Astrobiologist Looks for Life in Space

A rocket launches perfectly
no mistakes allowed.
A rover lands on another planet and starts exploring.
A satellite orbits Earth, sending back data every second.
Space is beautiful—but it's also very unforgiving.
If you like extreme challenges, careful planning, and technology that must work the first time, you might already be thinking like a
space engineer.

What They Do

Astrobiologists study life in the universe.
They ask questions like:

- What is life, really?
- Where can life survive?
- What conditions does life need?
- Could life exist beyond Earth?

They study:

- Tiny organisms that live in extreme places on Earth
- Mars, moons, and planets
- Chemicals linked to life
- Space environments

By studying life on Earth, they learn where life might exist elsewhere.

Where You See Their Work in Real Life

You see astrobiologists' work when:

- Space missions search for water on Mars
- Scientists study ice-covered moons
- News talks about "habitable planets"
- Rovers analyze soil for signs of life
- Telescopes study distant planets

Astrobiologists help guide where space missions look—and what they look for.

Cool Tools & Tech

Astrobiologists use tools like:

- Space probes and rovers
- Microscopes
- Chemical sensors
- Spectrometers (to study light and gases)
- Lab simulations of space conditions

Wow fact:

👉 Some organisms on Earth can survive extreme heat, cold, or radiation–helping scientists imagine life in space.

Future Problem They Help Solve

Astrobiologists help solve problems like:

- Finding out if life exists beyond Earth
- Understanding how life began
- Protecting Earth from harmful space microbes
- Preparing humans for long space missions
- Discovering how rare–or common–life might be

They help answer one of humanity's biggest questions:
Are we alone?

Try It Now: Extreme Life Thinker

You'll need:

- Paper
- Pencil

Steps:

1. Pick an extreme place on Earth (deep ocean, desert, ice).
2. Imagine what kind of life could survive there.
3. Now imagine a similar place on another planet.

That's how astrobiologists connect Earth life to space.

You'd Like This If You...

- Love space and science
- Enjoy big questions
- Like biology and chemistry
- Are curious about the unknown
- Dream about discovery

Why this matters to you:

Astrobiology helps humans understand life itself—and where it might exist beyond Earth.

Astrobiologists don't just search for aliens.

They search for answers about life.

Ocean Engineer
Underwater Tech and Structures

A robot explores the deep ocean where humans can't go.
Strong cables and pipes sit safely on the ocean floor.
Underwater machines help scientists study sea life and earthquakes.
Most of the ocean is still unexplored.
If you like engineering and the ocean—and enjoy solving problems in tough environments—you might already be thinking like an **ocean engineer**.

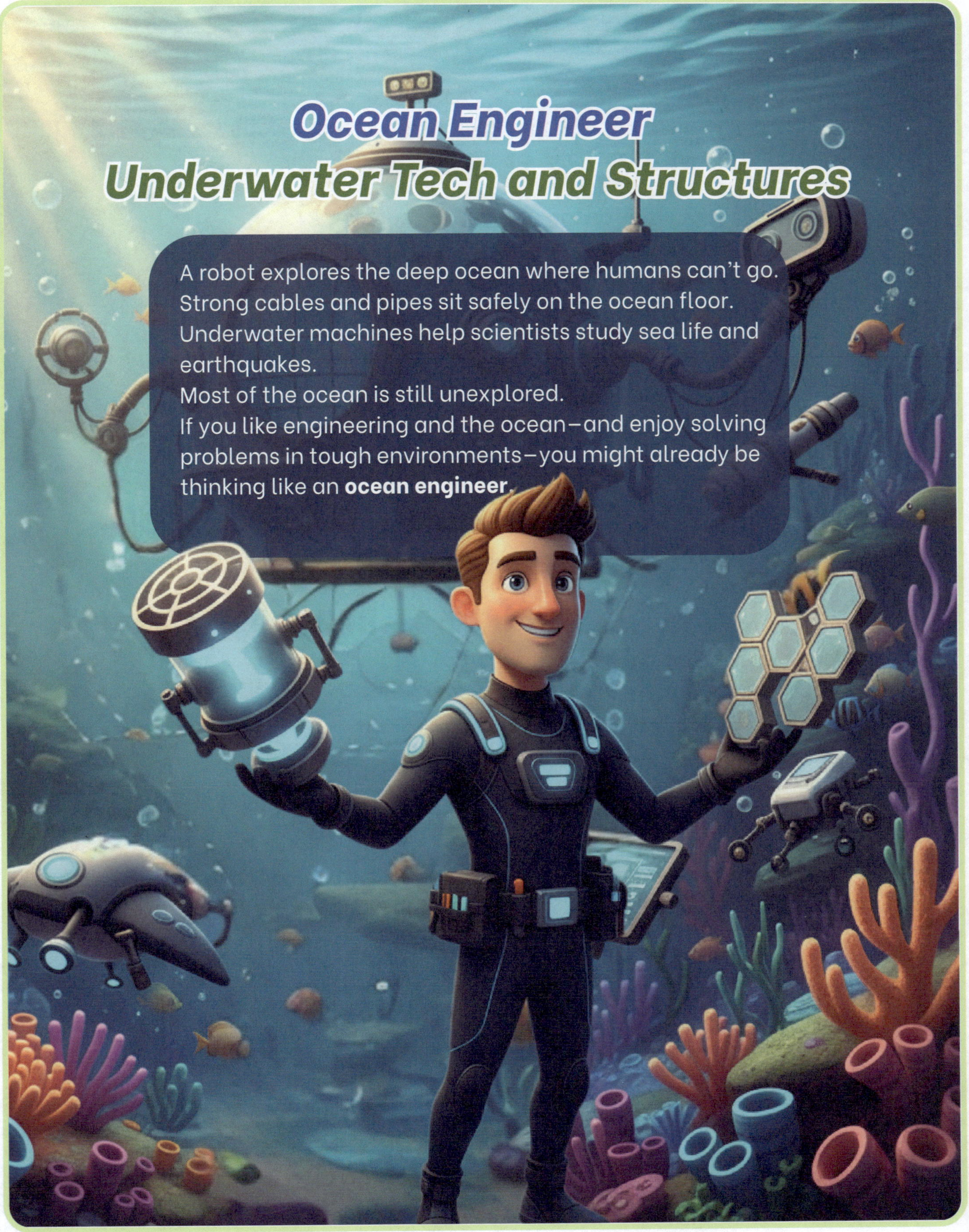

What They Do

Ocean engineers design technology that works underwater.
They build and improve:

- Underwater robots and vehicles
- Sensors that study ocean life and earthquakes
- Structures like underwater cables and platforms
- Equipment that can survive deep pressure

They think about:

- Water pressure
- Strong waves and currents
- Corrosion from saltwater
- How machines move underwater

The ocean is powerful—and ocean engineers design tools that can handle it.

Where You See Their Work in Real Life

You see ocean engineers' work when:

- Scientists explore deep-sea environments
- Internet cables run across oceans
- Oil and wind platforms operate offshore
- Underwater drones map the sea floor
- Tsunami sensors monitor ocean movement

Even your internet connection may rely on ocean engineering.
pace missions look—and what they look for.

Cool Tools & Tech

Ocean engineers use tools like:

- Underwater robots (ROVs and AUVs)
- Pressure-resistant materials
- Sonar systems
- Ocean sensors
- Computer simulations

Wow fact:

👉 At the deepest parts of the ocean, pressure can be over 1,000 times stronger than at the surface.

Future Problem They Help Solve

Ocean engineers help solve problems like:

- Exploring unknown parts of the ocean
- Protecting underwater ecosystems
- Studying earthquakes and tsunamis
- Building clean offshore energy
- Monitoring climate change in oceans

They help humans understand and respect the largest part of our planet.

Try It Now: Float or Sink Test

You'll need:

- *A bowl of water*
- *Small objects (coin, leaf, paperclip, plastic cap)*

Steps:

1. Place each object in water.
2. Observe which floats and which sinks.
3. Think about how shape and material matter.

Ocean engineers test buoyancy like this—just with bigger tools.

You'd Like This If You...

- Love the ocean
- Enjoy engineering challenges
- Like robots and machines
- Are curious about unexplored places
- Want to work with science and technology

Why this matters to you:

The ocean affects weather, food, and life on Earth—and ocean engineers help us understand it safely.

Ocean engineers don't just work underwater.

They help unlock the secrets of the deep.

Geoengineer / Hydrologist
Works with Earth & Water

Rain falls—but instead of flooding homes, it flows safely away.
A river provides water without drying up.
A city prepares for droughts before they happen.
Water shapes the Earth and life depends on it.
If you like nature, maps, and understanding how land and water work together, you might already be thinking like a **geoengineer or hydrologist.**

What They Do

Geoengineers and hydrologists study Earth, water, and how they interact. They focus on things like:

- Rivers, lakes, and groundwater
- Rainfall and drought
- Floods and erosion
- Soil and rock layers
- How humans affect water systems

Hydrologists mostly study water movement.
Geoengineers design solutions to manage Earth and water safely.

Together, they help protect communities and nature.

Where You See Their Work in Real Life

You see their work when:

- Flood control systems protect cities
- Dams and reservoirs manage the water supply
- Wells provide clean drinking water
- Farmers plan irrigation
- Cities prepare for droughts or heavy rain

If water reaches people safely—and not destructively—these experts helped plan it.

Cool Tools & Tech

Geoengineers and hydrologists use tools like:

- Water sensors to measure flow and quality
- Maps and satellite images
- Computer models that predict floods
- Soil and rock samples
- Weather and climate data

Wow fact:

👉 Much of Earth's fresh water is hidden underground—and scientists map it without ever seeing it.

Future Problem They Help Solve

They help solve problems like:

- Preventing floods and landslides
- Managing water during droughts
- Protecting drinking water sources
- Adapting to climate change
- Designing safe land use

As the weather becomes more extreme, their work becomes more important.

Try It Now: Water Path Finder

You'll need:

- Paper
- Pencil

Steps:

1. Draw a small hill and a house below it.
2. Draw where rainwater would flow.
3. Add channels or barriers to guide the water safely.

That's how water engineers begin designing solutions.

You'd Like This If You...

- Care about water and nature
- Enjoy maps and Earth science
- Like problem-solving outdoors
- Think ahead about safety
- Want to protect communities

Why this matters to you:

Water can give life—or cause damage. Understanding it helps protect both people and the planet.

Geoengineers and hydrologists don't fight nature.

They learn from it—and work with it.

STEM Entrepreneur
Starts Their Own STEM Company

You notice a problem no one has solved yet.
You have an idea that could help people.
Instead of waiting for someone else to fix it, you build something new.
That's what entrepreneurs do.
If you like creating ideas, taking risks, and turning science or technology into real solutions, you might already be thinking like a **STEM entrepreneur.**

What They Do

STEM entrepreneurs start companies using science, technology, engineering, or math.
They don't all invent new gadgets.
Some build:

- Apps that solve problems
- Clean energy solutions
- Medical tools
- Educational games
- Smart systems for cities, food, or health

They combine:

- STEM knowledge
- Creativity
- Problem-solving
- Leadership

Their job is to take an idea and turn it into something that actually works in the real world.

Where You See Their Work in Real Life

You see STEM entrepreneurs' work when:

- A new app makes life easier
- A startup creates eco-friendly products
- A small company grows into a big one
- New technology enters schools or hospitals
- Solutions appear that didn't exist before

Many of today's biggest companies started as one person's idea.
helped plan it.

Cool Tools & Tech

STEM entrepreneurs use tools like:

- Prototypes and early models
- Design software
- Coding platforms
- Business plans
- Feedback from real users

Wow fact:

👉 Many successful entrepreneurs failed several times before their idea worked.
Failure isn't the end—it's part of the process.

Future Problem They Help Solve

STEM entrepreneurs help solve problems like:

- Creating jobs
- Bringing new ideas to life
- Solving problems that governments or big companies miss
- Making technology more human-centered
- Turning science into solutions people can actually use

They often work on problems that don't even have names yet.

Try It Now: Be an Idea Starter

You'll need:

- Paper
- Pencil

Steps:

1. Write down one problem you notice in daily life.
2. Think of a STEM-based idea that could help.
3. Sketch or describe your solution.

That's how many companies begin—with one clear problem and one curious mind.

You'd Like This If You...

- Love coming up with ideas
- Enjoy building things from scratch
- Aren't afraid to try and fail
- Want to help people in new ways
- Like leading projects

Why this matters to you:

STEM entrepreneurship lets you shape your own path—and create solutions the world needs.

STEM entrepreneurs don't wait for the future.

They built it.

Designing YOUR STEM Path

You've just met 50 different STEM heroes.
That's a lot.
So if your brain feels full right now, that's normal.
You might feel:

- Super excited about a bunch of them
- Curious about a few
- Not interested in some at all

That's not a problem.
That's information.

Here's the most important thing to remember:

You do NOT have to choose one career. Not now. Not yet.
Real adults don't follow straight lines.

Many of them:

- Change careers
- Combine fields (like biology + coding, or art + engineering)
- Discover what they love slowly
- Try something... then try something else

STEM is not about picking the "right job."
It's about learning how you think.

Who STEM Is Really For

STEM is NOT only for:

- Kids who get perfect grades
- Kids who love math homework
- Kids who already "know what they want to be"

STEM is for kids who:

- Ask "why?" a lot
- Get deeply curious about one topic
- Like solving problems (even small ones)
- Notice when something feels broken or unfair
- Care about people, animals, or the planet

If that sounds like you—even a little—
you already belong here.

You Can Start Small (Like... Really Small)

You don't need special equipment.
You don't need a big plan.
Try this one-month STEM mini-adventure:

Week 1 – Explore
Pick 3 careers from this book.
Watch a kid-friendly video or read a short article about each one.
No pressure. Just curiosity.

Week 2 – Try
Choose one "Try It Now" activity from the book.
Do it at home or school.
Notice how it feels:

- Fun?
- Frustrating?
- Interesting?
- Boring?

All answers are okay.

Week 3 – Share
Tell someone:

- A friend
- A parent
- A teacher

Say:
"I learned about this career, and I liked this part..."
Talking helps your brain understand what matters to you.

Week 4 – Reflect
Ask yourself:

- What did I enjoy most?
- What did I not enjoy?
- What surprised me?
- What questions do I still have?

Write it down.
That notebook is your first STEM journal.
It's not homework.
It's your map of future-you.

The World Needs Your "Weird Mix"

You are not just one thing.
Maybe you love:

- Drawing and animals
- Sports and numbers
- Stories and technology
- Nature and building things

That mix is not random.
Some of the most important future jobs do not exist yet.
They will be created by people who:

- Combine skills
- Think differently
- Care deeply
- Notice problems others ignore
-

That could be you.

Your Job Right Now

Your job is not to know everything.
Your job is to:

- Stay curious
- Be kind
- Notice problems
- Try small experiments to fix them

That is what every STEM hero in this book actually does.

STEM STYLE QUIZ

Discover How YOUR Brain Likes to Work

This is not a test you can fail.

There are no wrong answers.

Just choose what feels most like you.

How to Answer

For each question, circle the letter that fits you best.

1. When something breaks, you usually...
A. Take it apart
B. Ask why it stopped working
C. Think of a better design
D. Think about who it affects

2. You feel proudest when you...
A. Build or fix something
B. Discover new information
C. Solve a tricky problem
D. Help someone

3. In a group project, you usually...
A. Build or set things up
B. Research facts
C. Plan the strategy
D. Make sure everyone feels included

4. Your favorite kind of challenge is...
A. Hands-on
B. Investigating
C. Logical
D. Real-world

5. If you had free time, you'd rather...
A. Create something
B. Learn something new
C. Solve puzzles
D. Improve something

6. When learning, you like it best when you can...
A. Try it yourself
B. Understand how it works
C. See patterns
D. Know why it matters

7. You're most curious about...
A. How things are built
B. How nature or the universe works
C. How systems make decisions
D. How to protect people or the planet

8. If you see a problem, you first think...
A. What tools do I need?
B. What's causing this?
C. What's the smartest solution?
D. Who is affected?

9. You enjoy school most when you are...
A. Making projects
B. Doing experiments
C. Solving challenges
D. Working with others

10. Which sounds most fun?
A. Building a robot
B. Studying animals or space
C. Designing a game strategy
D. Creating something that helps others

11. You learn best when...
A. You move or build
B. You observe closely
C. You think deeply
D. You connect ideas to real life

12. Your brain feels happiest when...
A. Things work
B. Things make sense
C. Things fit together logically
D. Things feel meaningful

13. If you could solve one problem, you'd choose...
A. Making tools better
B. Understanding a mystery
C. Improving a system
D. Helping people live better

14. You usually notice...
A. How objects are made
B. Patterns in nature
C. Patterns in numbers
D. How people feel

15. When a plan doesn't work, you...
A. Adjust and rebuild
B. Study what went wrong
C. Try a new strategy
D. Think about the impact

16. You feel confident when...
A. Using tools
B. Explaining facts
C. Solving puzzles
D. Supporting others

17. You'd rather be known as someone who...
A. Builds things
B. Discovers things
C. Figures things out
D. Makes a difference

18. You enjoy stories about people who...
A. Invent
B. Explore
C. Outsmart problems
D. Help the world

19. Your curiosity usually starts with...
A. "How do I make this?"
B. "Why does this happen?"
C. "What's the best way?"
D. "How does this help?"

20. If you could choose a superpower, it would be...
A. Building anything
B. Knowing how things work
C. Solving impossible puzzles
D. Making the world better

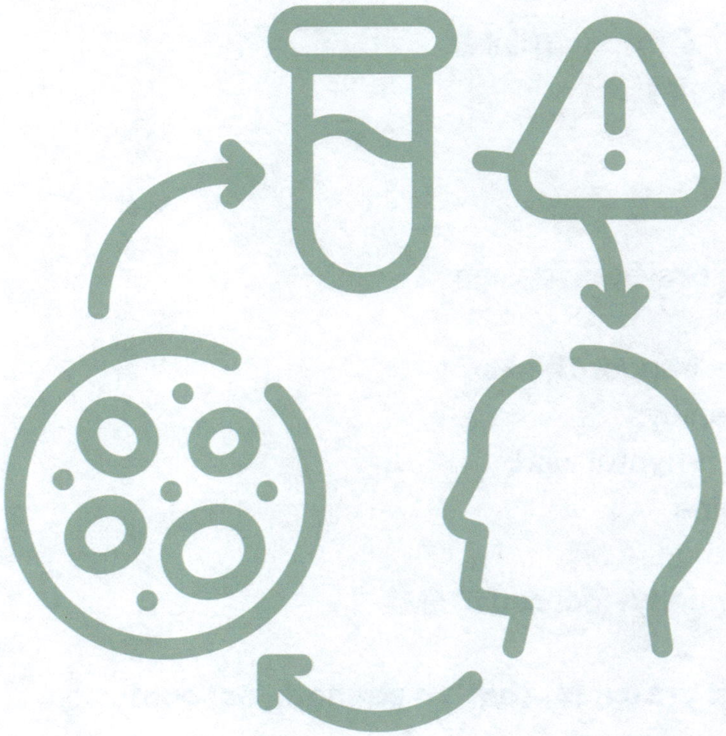

Your Results

Count how many times you chose A, B, C, and D.

Mostly A – Builder Thinker

You may enjoy:

- Engineering
- Robotics
- Inventing
- Mechanics

Mostly B – Explorer Thinker

You may enjoy:

- Science
- Biology
- Earth or space studies
- Research

Mostly C – Strategy Thinker

You may enjoy:

- Math
- Coding
- Data
- Game or system design

Mostly D – Impact Thinker

You may enjoy:

- Environmental work
- Medicine
- Education
- Community-focused STEM

Many kids are a mix. That's a strength, not confusion.

What This Quiz Really Means

This quiz doesn't tell you what job to pick.
It helps you notice:

- What excites you
- How your brain likes to work
- Which careers to explore first

Your answers can change—and that's okay.

STEM isn't a single path.

It's a toolbox.

And you're just starting to open it.

One Last Thing

You don't need a title yet.
You don't need a label.
You just need to keep asking questions.
Because the world isn't waiting for perfect answers.
It's waiting for curious thinkers who care.
And you're already becoming one.

So... which **STEM superjob** will you explore next?

My STEM Curiosity Reflection Sheet

A guide to discovering how you think

Name: ______________________________

Date: ______________________________

Book: So You Think You Might Be a... STEM Explorer

Chapters I Connected with Most

Circle or write the chapter numbers or names that caught your attention.

My Top 3:

1.
2.
3.

One chapter that didn't feel like me:

What Caught My Attention

(Choose One From Each Row)

I noticed I liked	Check one or more
• Solving puzzles or fixing problems	☐
• Building or making things	☐
• Asking why and testing ideas	☐
• Working with numbers or patterns	☐
• Helping people	☐
• Protecting the planet	☐
• Designing games, systems, or tools	☐
• Imagining new ideas or inventions	☐

My Brain at Work

Finish the sentences honestly.

There are no wrong answers.

- When something doesn't work, I usually feel:
- □ curious □ annoyed □ determined □ confused □ excited
- I enjoy learning best when I can:
- □ try things myself
- □ watch closely
- □ ask questions
- □ draw or design
- □ think quietly
- I feel most confident when I'm:

Skills I'm Already Using

Check the skills you think you already practice—even a little.

- □ Curiosity
- □ Patience
- □ Creativity
- □ Problem-solving
- □ Observation
- □ Logical thinking
- □ Caring about others
- □ Caring about the planet
- One skill I want to grow in more:

My Curiosity Clues

Answer at least one of the questions below.

- One problem I'd like to solve someday:
- Something I want to learn more about:
- One thing from this book that surprised me:

Reflection (Short & Powerful)

Circle one and explain why.

This book helped me understand:

□ how I think

□ what I'm curious about

□ what I enjoy learning

□ that I don't need all the answers yet

Because:

One Final Thought

Write one sentence to your future self:

"Keep being curious about..."

Made in United States
Cleveland, OH
05 November 2026

47768323R00121